Biennial State-of-the-Art Sensors Technology in Australia 2019-2020

Biennial State-of-the-Art Sensors Technology in Australia 2019-2020

Editors

Edith Chow
Rona Chandrawati

MDPI • Basel • Beijing • Wuhan • Barcelona • Belgrade • Manchester • Tokyo • Cluj • Tianjin

Editors
Edith Chow
Aperture
Australia

Rona Chandrawati
School of Chemical Engineering,
University of New South Wales
Australia

Editorial Office
MDPI
St. Alban-Anlage 66
4052 Basel, Switzerland

This is a reprint of articles from the Special Issue published online in the open access journal *Sensors* (ISSN 1424-8220) (available at: https://www.mdpi.com/journal/sensors/special_issues/State-of-the-Art_Sensors_Australia_2019-2020).

For citation purposes, cite each article independently as indicated on the article page online and as indicated below:

LastName, A.A.; LastName, B.B.; LastName, C.C. Article Title. *Journal Name* **Year**, *Volume Number*, Page Range.

ISBN 978-3-0365-0708-8 (Hbk)
ISBN 978-3-0365-0709-5 (PDF)

© 2021 by the authors. Articles in this book are Open Access and distributed under the Creative Commons Attribution (CC BY) license, which allows users to download, copy and build upon published articles, as long as the author and publisher are properly credited, which ensures maximum dissemination and a wider impact of our publications.
The book as a whole is distributed by MDPI under the terms and conditions of the Creative Commons license CC BY-NC-ND.

Contents

About the Editors . . . . **vii**

Preface to "Biennial State-of-the-Art Sensors Technology in Australia 2019-2020" . . . . **ix**

Elicia L. S. Wong, Khuong Q. Vuong and Edith Chow
Nanozymes for Environmental Pollutant Monitoring and Remediation
Reprinted from: *Sensors* **2021**, *21*, 408, doi:10.3390/s21020408 . . . . **1**

Felix Weihs, Alisha Anderson, Stephen Trowell and Karine Caron
Resonance Energy Transfer-Based Biosensors for Point-of-Need Diagnosis—Progress and Perspectives
Reprinted from: *Sensors* **2021**, *21*, 660, doi:10.3390/s21020660 . . . . **49**

Xinyue Yao, Mikko Vepsäläinen, Fabio Isa, Phil Martin, Paul Munroe and Avi Bendavid
Advanced RuO_2 Thin Films for pH Sensing Application
Reprinted from: *Sensors* **2020**, *20*, 6432, doi:10.3390/s20226432 . . . . **67**

Alessio Scalisi, Daniele Pelliccia and Mark Glenn O'Connell
Maturity Prediction in Yellow Peach (*Prunus persica* L.) Cultivars Using a Fluorescence Spectrometer
Reprinted from: *Sensors* **2020**, *20*, 6555, doi:10.3390/s20226555 . . . . **81**

Hamideh Keshavarzi, Caroline Lee, Mark Johnson, David Abbott, Wei Ni and Dana L. M. Campbell
Validation of Real-Time Kinematic (RTK) Devices on Sheep to Detect Grazing Movement Leaders and Social Networks in Merino Ewes
Reprinted from: *Sensors* **2021**, *21*, 924, doi:10.3390/s21030924 . . . . **99**

Rhiannon Alder, Jungmi Hong, Edith Chow, Jinghua Fang, Fabio Isa, Bryony Ashford, Christophe Comte, Avi Bendavid, Linda Xiao, Kostya (Ken) Ostrikov, Shanlin Fu and Anthony B. Murphy
Application of Plasma-Printed Paper-Based SERS Substrate for Cocaine Detection
Reprinted from: *Sensors* **2021**, *21*, 810, doi:10.3390/s21030810 . . . . **119**

About the Editors

Edith Chow completed a PhD in chemistry from the University of New South Wales in 2006. From 2006 to 2020 she worked at CSIRO, Australia's national science agency, where she co-invented a gold nanoparticle chemiresistor sensor technology that could be used to detect trace organics. Her sensor technology attracted over 10M USD in funding, for which she partnered with the government, international agencies, and small and medium enterprises to tackle the problems faced by the environmental, agricultural, food and healthcare sectors. In 2020, she co-founded Aperture to focus on the commercialisation of nanomaterial-based sensors. Edith is the author of over 40 peer-reviewed publications (h-index 21), two patents and one book chapter on chemical sensing, nanomaterials and electrochemistry. She is a Fellow of the Royal Australian Chemical Institute and Royal Society of New South Wales.

Rona Chandrawati is a Scientia Associate Professor and National Health and Medical Research Council Emerging Leadership Fellow at the School of Chemical Engineering, The University of New South Wales, Australia. She obtained her Ph.D. from the Department of Chemical and Biomolecular Engineering at The University of Melbourne in 2012. She was then a Marie Curie Fellow at Imperial College London, before returning to Australia in 2015 to establish her research group (Nanotechnology for Food and Medicine Laboratory). Over the last 8 years, Rona has successfully attracted over AUD 4M in competitive funding as a Lead Chief Investigator and AUD10M in funding as a co-Chief Investigator. She is the author of over 70 peer-reviewed publications with >3000 citations (h-index 24). Rona was named in Australia's Most Innovative Engineers 2020 by Engineers Australia. Her research interests include the development of nanoparticles, polymers, natural enzymes and enzyme mimics for drug delivery and sensing.

Preface to "Biennial State-of-the-Art Sensors Technology in Australia 2019-2020"

Recent developments in novel materials, sensing principles and signal processing are paving the way for new types of analytical sensors, offering an improved sensing performance and practicality. Such advancements will provide accurate and real-time analytical information in the environmental, food, and healthcare sectors to better guide decision-making. For sensors to be commercially viable, the miniaturisation and integration of components for rapid, automated, in-field detection and diagnosis need to be considered. This book highlights state-of-the-art sensors technology in Australia through original contributions and reviews.

Edith Chow, Rona Chandrawati
Editors

Review

Nanozymes for Environmental Pollutant Monitoring and Remediation

Elicia L. S. Wong, Khuong Q. Vuong and Edith Chow *

Aperture, Ryde, NSW 2112, Australia; elicia.wong@apertureteam.com (E.L.S.W.); khuong.vuong@apertureteam.com (K.Q.V.)
* Correspondence: edith.chow@apertureteam.com

Received: 12 December 2020; Accepted: 6 January 2021; Published: 8 January 2021

Abstract: Nanozymes are advanced nanomaterials which mimic natural enzymes by exhibiting enzyme-like properties. As nanozymes offer better structural stability over their respective natural enzymes, they are ideal candidates for real-time and/or remote environmental pollutant monitoring and remediation. In this review, we classify nanozymes into four types depending on their enzyme-mimicking behaviour (active metal centre mimic, functional mimic, nanocomposite or 3D structural mimic) and offer mechanistic insights into the nature of their catalytic activity. Following this, we discuss the current environmental translation of nanozymes into a powerful sensing or remediation tool through inventive nano-architectural design of nanozymes and their transduction methodologies. Here, we focus on recent developments in nanozymes for the detection of heavy metal ions, pesticides and other organic pollutants, emphasising optical methods and a few electrochemical techniques. Strategies to remediate persistent organic pollutants such as pesticides, phenols, antibiotics and textile dyes are included. We conclude with a discussion on the practical deployment of these nanozymes in terms of their effectiveness, reusability, real-time in-field application, commercial production and regulatory considerations.

Keywords: degradation; detection; enzyme; heavy metal; nanomaterial; nanoparticle; peroxidase; pesticide; pollution; sensor

1. Introduction

1.1. General Introduction to Nanozymes

Nanozymes are advanced nanomaterials which possess unique physicochemical properties with the precise structural fabrication capability to mimic intrinsic biologically relevant reactions. Specifically, nanozymes mimic natural enzymes and exhibit enzyme-like properties. The enzymatic catalytic reactions are highly effective, with reactions occurring rapidly even under mild conditions, and more importantly, such reactions are also highly selective. The high efficiency and selectivity are immensely desirable properties for sensing and monitoring applications. However, natural enzymes including proteins suffer from limitations such as low thermostability and narrow pH window, which will denature the enzymes and greatly reduce and inhibit their enzymatic activities. Low thermostability also places stringent requirements on the storage, transportation and handling of natural enzymes, which can be labour- and infrastructure-intensive for the users. Susceptible denaturation adds complexity to the interpretation of sensing and monitoring outputs, which may yield a false positive/negative outcome. From this perspective, nanozymes address these limitations by offering high structural durability and stability, while maintaining the desirable catalytic activities.

By incorporating the unique physicochemical properties and enzyme-like activities, nanozymes exhibit promising applications in different fields such as the biomedical sector (in vivo diagnostics/and therapeutics) and the environmental sector (detection and remediation of inorganic and organic pollutants). The biomedical and clinical translation of nanozymes have been extensively reviewed [1–4], from their applications in immunoassays [5–10] to cancer diagnostics and therapeutics [11–15]. As nanozymes offer better structural stability over their respective natural enzymes, with wider physical (e.g., temperature) and chemical (e.g., pH) operational windows, they are ideal candidates for real-time and/or remote environmental monitoring and remediation. This is especially so given the challenging and unpredictable nature of the outdoor environment (compared to a more physiologically stable in vivo or in vitro environment which has a more defined and narrower operational window). In this review, the current environmental translation of nanozymes into a powerful sensing or remediation tool through inventive (i) nano-architectural design of nanozymes and (ii) transduction methodologies will be reviewed, as well as the practical deployment of these nanozymes in terms of their functionality and recyclability.

1.2. Types of Nanozymes

Over the last few decades, various types of nanomaterials have been reported to have intrinsic enzyme-like activities [16,17]. Natural enzymes exhibit intrinsic catalytic ability, usually at a single active site, to catalyse a specific chemical transformation [18,19]. Since nanozymes lack such an active site, different strategies have been devised to enhance the catalytic properties of these nanomaterials, enabling them to selectively and effectively react with target molecules. In this review, we categorise nanozymes into four types based on their mode of natural enzyme-mimicking behaviour (Figure 1).

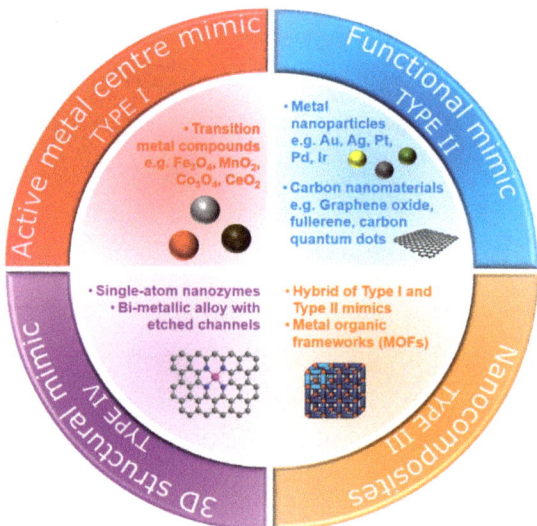

Figure 1. Types of nanozymes based on their mode of natural enzyme-mimicking behaviour.

One strategy utilises constructed metal sites (type I nanozymes), such as metal oxides or metal sulphides, to mimic the metal catalytic active site found in metalloenzymes [20]. An early example of such nanozymes is iron oxide (Fe_3O_4) nanoparticles, as reported by Gao et al. [6], which exhibit peroxidase-like activity similar to the natural horseradish peroxidase (HRP) enzyme. Peroxidases catalyse the oxidation of a chromogenic substrate, such as 3,3′,5,5′-tetramethylbenzidine (TMB), 2,2′-azino-bis(3-ethylbenzothiazoline-6-sulfonic acid) (ABTS), *o*-phenylenediamine dihydrochloride (OPD), in the presence of hydrogen peroxide, H_2O_2. The smallest Fe_3O_4 nanoparticles (out of 30, 150 and 300 nm) were found to have the highest catalytic activity, which showed that the high surface area of the Fe_3O_4 nanoparticles was responsible for the peroxidase-like activity. Gao et al. [6] postulated that the active surface ferrous and ferric ions in the nanoparticles were the key components that enabled this catalytic activity, mimicking the iron–heme binding site in HRP. Cerium oxide (CeO_2) nanoparticles also utilise their metal as a nanozyme due to its similarity in structure and biochemistry to the iron ion, particularly in binding to proteins [21]. In fact, CeO_2 nanoparticles are multifunctional catalysts whereby they exhibit catalase-like (breakdown of H_2O_2 into O_2 and H_2O) and superoxidase-like (dismutation of $O_2^{\bullet-}$ into O_2 and H_2O_2) activities in addition to bearing peroxidase-like activity. The multifunctional catalytic behaviour arises from the coexistence of Ce(III) and Ce(IV) oxidation states. The switch between the III/IV valence resembles the mechanism of redox enzymes, which use metals as cofactors to catalyse a range of reversible redox reactions [16]. As such, reactions comprising redox cycles between Ce(III) and Ce(IV) oxidation states make it possible for CeO_2 nanoparticles to react catalytically with oxygen radicals and hydrogen peroxide, thus mimicking the function of two key antioxidant enzymes, namely superoxide and catalase [22]. Similar to the Fe_3O_4 nanoparticles, the enzyme-mimicking behaviours of CeO_2 are also size-dependent, whereby the smaller CeO_2 (5 nm) have superior catalytic activity to that of larger (28 nm) nanoparticles [23]. This is further exemplification that the catalytic activities of the nanozymes are dependent on the amount of catalytically active atoms exposed on the surface (where the catalytic mechanism is related to changes in metal valence), which is usually inversely proportional to the diameter of the nanoparticles.

Other common metal–heme centres found in metalloenzymes include copper, cobalt and manganese ions (Table 1). As such, nanoparticles synthesised from cobalt oxide (Co_3O_4) [24], cobalt sulphide (Co_9S_8) [25], copper oxide (CuO) [26] and manganese dioxide (MnO_2) [27,28] are also known to possess either peroxidase, oxidase and/or catalase-mimicking activities.

Table 1. Examples of metalloenzymes.

Metal Centre	Enzymes
Zinc	Carbonic anhydrase, alcohol dehydrogenase, organophosphate hydrolase
Iron	Catalase, peroxidase, cytochrome oxidase
Manganese	Enolase, hexokinase
Copper	Tyrosinase, lysyl oxidase, laccase
Cobalt	Dipeptidase

While the type I metal compound nanoparticles are those that mimic the metal–heme redox centre of metalloenzymes, other metal nanoparticles that catalyse the same reactions as natural enzymes are classified as type II nanozymes. These metal nanoparticles are synthesised from metals which are known to exhibit intrinsic catalytic behaviour for various heterogeneous reactions. Such metals include gold, silver, platinum, palladium and iridium [29–31]. Rossi and co-workers reported that gold nanoparticles, under controlled conditions (unprotected "naked" nanoparticles, 3.6 nm diameter, in the presence of excess glucose) could initially catalyse reactions similar to glucose oxidase and thus serve as a mimic for glucose oxidase [32]. Moreover, gold nanoparticles also showed peroxidase-mimicking activity [33,34]. Chen and co-workers [33] demonstrated that unmodified gold nanoparticles had significantly higher catalytic behaviours towards peroxidase substrates, which indicated that the superficial gold atoms were the key component to the observed peroxidase-like activity. Modified gold nanoparticles with different surface charges (positive or negative) also exhibited peroxidase-mimicking activity [34]. In fact, it was found that their enzyme-mimetic activities could be modulated by changing the pH of the environment (i.e., pH-switchable). Gao and co-workers [35] demonstrated that gold, silver, platinum and palladium nanomaterials exhibited peroxidase-like activities at acidic pH and catalase-like activities at basic pH [35]. The pH-switchable phenomenon was further investigated by Nie and co-workers using 1–2-nm platinum nanoparticles, which showed that the catalase-like activity was evident under basic conditions while the peroxidase-like activity was more dominant under acidic conditions [36]. The catalytic mechanism of the metal nanoparticles is different from the metal compound-based nanozymes and is related to the adsorption, activation and electron transfer of substrate (e.g., TMB, ABTS or OPD) on metal surfaces rather than a change in the metal valence of the nanomaterial.

Another intriguing aspect of metal-based nanozymes is that they can form alloys with different elemental compositions [37]. By combining the independent electronic characteristics of two metals, bimetallic nanoparticles can further exhibit unique properties through the synergetic effect of the two metals [38]. Thus, this makes it feasible to tailor the enzyme-mimicking activities by adjusting alloy compositions, classified here as type III nanozymes. In one example, He et al. showed that for Ag-M (M = Au, Pd, Pt) bimetallic alloy-based peroxidase nanozymes, the efficiency of the catalytic activity could be tuned by gradually changing the ratio of the two metals [39]. They suggested that the composition-dependent activity was from the electronic structure due to alloying. In another example, further enhancement of the multi-enzymatic activities was demonstrated by Yin, Wu and co-workers [40] using Au-Pt bimetallic nanoparticles by controlling the Pt and Au molar ratio to exhibit oxidase, peroxidase and catalase-like activities. An enhanced peroxidase-like activity of Ir-Pd nanocubes was obtained by depositing an Ir atomic layer on the surface of Pd nanocubes [41]. It was postulated that the adsorption energy of the Ir-Pd(100) surface was larger than that of the Pd(100), making it more energy-efficient to dissociate hydrogen peroxides into hydroxyl radicals.

Non-metallic nanozymes such as carbon-based nanomaterials, including fullerene and their derivatives, carbon quantum dots, carbon nanotubes and graphene oxide, are showing great promise with enzyme-mimicking capability owing to their intrinsic catalytic properties. The peroxidase, catalase and oxidase-like activities have all been reported [42–47]. In one example, Shi et al. [44] reported that carbon quantum dots exhibited peroxidase-like catalytic activity. It was concluded that the catalytic mechanism came from an increase in the electron density and mobility in the carbon quantum dots acting as effective catalytically active sites. Qu and co-workers [45] also reported that carboxyl-modified graphene oxide exhibited peroxidase-like activity, with electron transfer occurring from the top of the valence band of graphene to the lowest unoccupied molecular orbital of hydrogen peroxide. To further lower this band gap and improve the peroxidase-like behaviour, Kim et al. [46] co-doped the graphene oxide with nitrogen and boron and demonstrated a much higher catalytic behaviour than undoped graphene oxide. Besides peroxidase-like behaviour, the catalase-like behaviour was reported by Ren et al. [47] using graphene oxide quantum dots.

Metal–carbon nanocomposites have also been investigated as a strategy to further improve the catalytic activities of carbon nanozymes [48–51]. An oxidase-like nanozyme, catalysing an oxidation–reduction reaction involving oxygen as an electron acceptor, was constructed using a metal–carbon nanocomposite hybrid through doping a N-rich porous carbon with Fe nanoparticles [51]. The group suggested that the N-doped porous carbon acted as the binding sites to trap and transfer O_2 molecules to catalytic sites and subsequently catalysed their redox reaction with the Fe nanoparticles. In another example, Guo, Zhang and co-workers [48] integrated graphene quantum dots with Fe_3O_4 nanoparticles and demonstrated superior peroxidase-like activities compared to individual graphene quantum dots and Fe_3O_4 nanoparticles. This superiority was attained from the synergistic interactions between graphene quantum dots and the Fe_3O_4 nanoparticles. Compared to the native HRP, this nanocomposite showed comparable, if not better, removal efficiencies for some phenolic compounds from aqueous solution, rendering it useful for industrial wastewater treatment [48].

All metal-, metal-compound- and carbon-based nanozymes rely on the high surface area, enabled either through small particle size (of the order of tens of nanometres for metal- or metal-compound-based nanozymes) or porous structure (carbon-based nanozymes) to maximise the exposure of the catalytically active atoms. Metal–organic frameworks (MOFs) which consist of metal ions as nodes and organic ligands as linkers also have highly porous structures that can be utilised as nanozymes. In this construct, the transition metal nodes containing the MOFs themselves can act as biomimetic catalysts, while the high porosity structure created by the metal–organic linkers can serve as the binding sites for the substrates. Their tuneable pore sizes, highly specific surface areas and exposed active sites provide MOFs with high catalytic efficiency [52]. In one example, Li and co-workers [53] demonstrated the use of a nanosized MOF, Fe-MIL-88NH$_2$, as a peroxidase mimic. The catalytic mechanism was proposed as follows: hydrogen peroxide was adsorbed onto the surface or into the mesopores of Fe-MIL-88NH$_2$, and the hydrogen peroxide was decomposed into hydroxyl radicals by iron. Other than the Fe-MOF, Cu-MOF [54], Ni-MOF [55], Pt-MOF [56] and Co/2Fe MOF [57] are also known to exhibit peroxidase-like behaviours. The bimetallic-MOF, Co/2Fe-MOF, exhibited dual enzymatic activities, peroxidase and oxidase. Additionally, Min, Chu and co-workers showed that CeO$_2$-MOF acted as a hydrolase mimic (breakage of a chemical bond using water) to remove a phosphate group, PO$_4^{3-}$, from phosphopeptides [58]. Prussian Blue nanoparticles are an analogue of MOFs which can simultaneously behave as multienzyme mimics (peroxidase, superoxide dismutase and catalase-like activities) and were used effectively as a scavenger for reactive oxygen species [59].

Although the aforementioned metal-compound-, metal-, carbon-, nanocomposite- and MOF-based nanomaterials show promising enzyme-mimicking abilities, achieving the same level of binding affinity and specificity as natural enzymes remains a challenge. The limiting factors include (i) the density of the catalytically active surface ions (such as the metal ions) and functional groups (such as the carboxyl group in the carbon nanomaterials), and (ii) the efficiency of the catalytic mechanism. It has been demonstrated that nanozymes with a low density of active sites show much lower catalytic activities [60]. Additionally, the elemental composition and facet structure of these nanozymes cause the catalytic mechanism of nanozymes to be different and are usually more intricate than natural enzymes [37,61]. These limitations constrain the extensive applications of these standard nanozymes. Consequently, new strategies have emerged to mitigate these constraints through spatial or three-dimensional structural mimicking of the active sites of natural enzymes [62,63]. These structural mimics can be achieved by mimicking the geometry of pre-existing metal binding centres, the binding sites at the peripheral or the confined and empty space at the centre of natural enzymes (type IV nanozymes).

Single-atom nanozymes resemble spatial structures to mimic the electronic, geometric and chemical structure of the pre-existing metal binding centre of metalloenzymes. For example, the FeN$_4$ in iron-based single-atom nanozymes mimic the active sites of oxymyoglobin, HRP and cytochrome P450 enzymes which contain a single heme Fe with a proximal ligand (Figure 2) [64,65]. In particular, Huang, Zhu and co-workers reported that densely isolated FeN$_4$ single-atom nanozymes exhibited outstanding peroxidase-like activities [66]. Both their experimental and theoretical analyses showed that FeN$_4$ led to strong adsorption of hydrogen peroxide, weakened the bonding between the single Fe atom and the two adsorbed hydroxyl groups and lowered the energy barrier for the formation of hydroxyl radicals to boost the peroxidase-like activities. Additionally, other metals such as cobalt and zinc could also be used to create CoN$_4$, and ZnN$_4$ single-atom nanozymes that exhibited peroxidase-like activities [66]. The addition of an axial N coordination to form FeN$_5$ single-atom nanozymes enhanced the oxidase-like behaviour of the Fe based single-atom nanozymes [62]. The FeN$_5$ structure had the most adsorption energy, by promoting strong oxygen adsorption that led to weakening of the O–O bond. Wang, Dong and co-workers [62] postulated that the weakening of the O–O bond was a result of the electron donor via the electron push effect of the axial-coordinated N in the FeN$_5$ single-atom nanozyme.

Other structure-mimicking strategies have been achieved through the creation of binding sites, such as nano-channels at the nanozymes, by resembling the binding sites of natural enzymes either at the periphery or in the centre (usually a confined empty cavity); see Figure 3. In a pioneering example, Schuhmann, Tilley, Gooding and co-workers designed a nanoparticle that mimicked the 3D architecture of a natural enzyme by using surfactant-covered PtNi bimetallic nanoparticles [63,67]. In their design, the surfactant-covered PtNi particles were selectively etched to create nano-channels that were specific for catalysing the reduction reaction of oxygen. The group reported that the oxygen reduction reaction activity normalised by the electrochemically active surface area was enhanced by a factor of 3.3 for the nanozymes compared to the unetched PtNi nanoparticles.

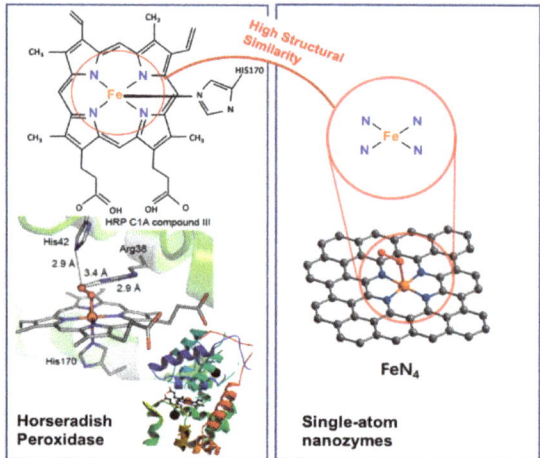

Figure 2. A comparison of the HRP enzyme and FeN$_4$ single-atom nanozyme showing the high structural similarity of the single-atom nanozyme to the active centre (iron-heme group) of HRP. (Adapted with permission from [64]. Copyright © 2017 Wiley-VCH Verlag GmbH & Co. KGaA, Weinheim, Germany).

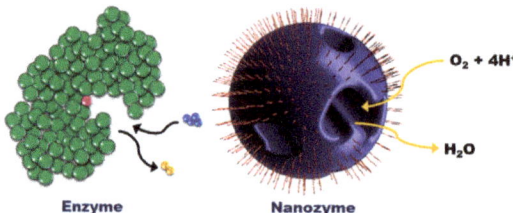

Figure 3. Illustration of a nanozyme as a 3D geometric architectural mimic of an enzyme. (Reprinted with permission from [63]. Copyright © 2018, American Chemical Society).

1.3. Catalytic Mechanisms

The most important advantage of nanozymes is their size-/composition-dependent activity, which enables the architectural design of nanomaterials with a broad range of catalytic activities by varying the shape, structure and composition. These nanozymes share certain similarities, such as that they need to be within a certain range of size, shape and surface charge to enable them to mimic natural enzymes. It is useful to acquire a basic insight into how these factors affect the catalytic performance of nanozymes. Thus, a fundamental understanding of the catalytic mechanisms behind the enzymatic-like activities is critical for further creative design of nanozymes with improved catalytic performance. In this section, we summarise the general mechanisms and analyse the catalytic mechanisms according to the types of nanozymes as introduced in the previous section.

Natural enzymes are internationally classified into six classes: oxidoreductases, transferases, hydrolases, lyases, isomerases and ligases. The majority of nanozymes are known to mimic the catalytic activities of oxidoreductases and hydrolases (Figure 4). Oxidoreductases catalyse oxidations and reductions in which hydrogen or oxygen atoms or electrons are transferred between molecules. Natural enzymes such as oxidases (including laccases), superoxide dismutases, peroxidases and catalases belong to the oxidoreductase family. Hydrolases catalyse the hydrolysis of various bonds and include enzymes such as phosphatase, nuclease, protease and peptidase.

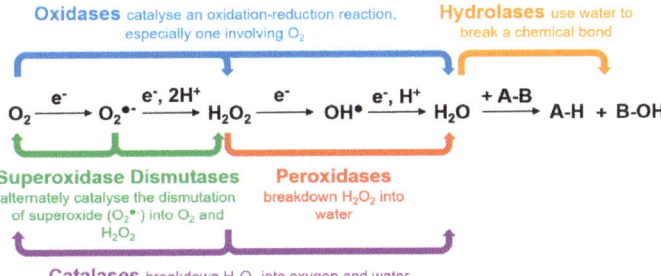

Figure 4. Schematic of the major reactions catalysed by oxidoreductases (oxidases, superoxide dismutases, peroxidases and catalases) and hydrolases.

The mode of catalytic reactions of natural enzymes involves two types of mechanisms, namely chemical and binding mechanisms. Chemical mechanisms include (i) acid-base catalysis (reactions involving H^+ and OH^- and are pH-dependent), (ii) covalent catalysis (reactions involving formation of a transient covalent bond) and (iii) metal ion catalysis (reactions involving redox changes and stabilisation of charges). Binding mechanisms involve (iv) proximity/orientation-assisted catalysis, (v) transition state stabilisation-assisted catalysis, and vi) electrostatic catalysis. Natural enzymes use one or a combination of these actions to catalyse a chemical transformation. Understandably, nanozymes also employ one or more of these mechanisms to mimic natural enzymes.

Metal compound-based nanozymes rely on changes in metal valence to catalyse the redox reaction, and this usually involves the use of transition metal elements which have variable oxidation states. These are also the metals commonly found in the metal–heme centre of metalloenzymes as well. Metal-compound-based nanozymes include Fe_3O_4, Co_3O_4, CeO_2, CuO and Mn_3O_4, in which the metal elements can be converted between variable valence states, making them promising nanozyme candidates. Using Fe_3O_4 nanoparticles as an example, Smirnov and co-workers [68] reported that the peroxidase-like activity of iron oxide originated mainly from the interaction of hydrogen peroxide with the ferrous ions on the surface of the nanoparticles (rather than from the dissolution of metal ions from the nanoparticles). These may follow Fenton reactions [69], which can be written as Equations (1)–(3). Other transition metals (M) follow similar Fenton reactions comprising redox cycles between $M(n)$ and $M(n+1)$ oxidation states.

$$Fe^{3+} + H_2O_2 \rightarrow FeOOH^{2+} + H^+ \quad (1)$$

$$FeOOH^{2+} \rightarrow Fe^{2+} + HO_2^\bullet \quad (2)$$

$$Fe^{2+} + H_2O_2 \rightarrow Fe^{3+} + OH^- + HO^\bullet \quad (3)$$

The catalytic mechanism of metal nanoparticles differs from metal-compound-based nanozymes and is related to the adsorption, activation and electron transfer of the substrate (e.g., TMB, ABTS or OPD) on metal surfaces. As mentioned in the previous section, metal-based nanozymes exhibit intrinsic pH-switchable peroxidase and catalase-like activities [35]. The pH-switchable ability arises from the acid-base type of catalysis of the metal-based nanozymes, which consists of the adsorption and decomposition of hydrogen peroxide under different pH conditions. Adsorption of hydrogen peroxide first occurs on the metal surface to initiate the catalytic reaction; then, the adsorbed hydrogen peroxide undergoes two different decomposition pathways depending on the pH of the micro-environment [35]. Under acidic conditions, hydrogen peroxide follows a base-like decomposition pathway to exhibit peroxidase-like activity, whereas under basic conditions, it follows an acid-like decomposition pathway exhibiting catalase-like activity. Figure 5 illustrates the decomposition mechanisms under acidic and basic conditions. Similar to the peroxidase and catalase-like activities, the catalytic mechanism for oxidase-like behaviour was demonstrated by Wu, Gao and co-workers [37] using density functional theory to involve the adsorption of oxygen to the metal surface, followed by the dissociation of oxygen. The proposed mechanism for the oxidase-like activity is shown in Equations (4) and (5). The same catalysis mechanisms as shown in Figure 5 and Equations (4) and (5) were also applicable to bimetallic [35,37,40,70,71] and MOF-based [72] nanozymes.

$$O_2 = 2O^* \tag{4}$$

$$O^* + S \rightarrow H_2O^* + S_{ox} \tag{5}$$

* is used to indicate species adsorbed on metal surfaces. S is a chromogenic substrate; S_{ox} is the oxidised chromogenic substrate.

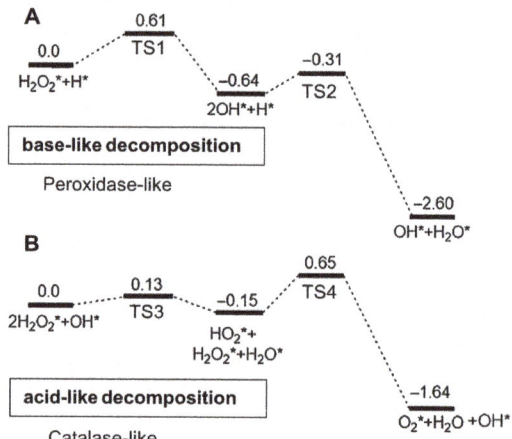

Figure 5. Calculated reaction energy profiles (unit: eV) for hydrogen peroxide decomposition on an Au(111) surface in (**A**) acidic and (**B**) basic conditions. Asterisk (*) is used to indicate species adsorbed on the metal surface. TS stands for transition state. (Adapted with permission from [35]. Copyright © 2015, Elsevier Ltd. All rights reserved).

Unlike metal-compound- and metal-based nanozymes, the catalytic mechanism of carbon-based nanozymes is not as well documented. However, it follows a catalytic route as for the metal-based nanozymes involving the adsorption, activation and electron transfer of the substrate at carbon surfaces. Qu and co-workers [43,73] completed a comprehensive study of the mechanism of the peroxidase-like activity of graphene oxide quantum dots, which stemmed from their ability to catalyse the decomposition of hydrogen peroxide and generate highly reactive hydroxyl radicals. These catalytic reactions occur at two sites, reactive and substrate-binding sites: the functional groups –C=O act as the catalytically active sites for converting hydrogen peroxide to hydroxyl radicals, and the O=C–O– groups serve as the substrate-binding sites for hydrogen peroxide (Figure 6). In a separate study, Zhao et al. [74] also reported a similar catalytic mechanism where the hydrogen peroxide attacked the –C=O group and H_2O_2 decomposed to form a hydroxyl radical.

For nanozymes that rely on structural analogy with the natural enzyme as the strategy to achieve the desired enzyme-mimicking properties, the high structural similarity maximises the binding energy of the substrate and lowers the activation energy required to initiate the catalytic reaction. Single-atom dispersion (for single-atom nanozymes) also generates the most sufficient size effect and provides maximum surface-active site exposure [75] because the catalytic mechanism depends primarily on the steric configuration of active centres instead of the size or structure of the nanoparticles as seen in the metal- and metal-compound-based nanozymes [62]. Using single-atom nanozymes with carbon nanoframe-confined FeN_5 active centres, Wang, Dong and co-workers [62] reported both experimental and theoretical studies of the catalytic mechanism of their oxidase-like activity. The sterically configured metal active centres (FeN_5) have a strong O_2 adsorption to the surface, weakening the O–O bond and giving rise to a larger extent of O–O bond elongation, thereby promoting the oxidase-like activity. Additionally, Xu et al. [76] also reported a similar catalytic mechanism for the peroxidase-like activity of single-atom nanozymes with the adsorption of hydrogen peroxide to the sterically configured active centres and the decomposition of hydrogen peroxide into reactive radicals.

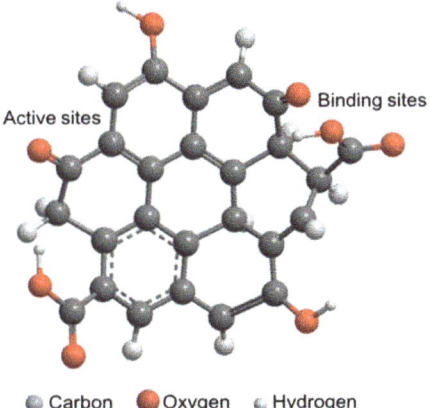

Figure 6. The catalytically active sites (for hydrogen peroxide) and binding sites (for the chromogenic substrate) of graphene oxide quantum dots which exhibit peroxidase-like activity. (Adapted with permission from [73]. Copyright © 2015 WILEY-VCH Verlag GmbH & Co. KGaA, Weinheim, Germany).

Instead of using the single-atom dispersion approach to maximise the surface-active site exposure, Calle-Vallejo et al. [77] introduced the concept of "coordination–activity plots" to predict the geometric structure of the optimal active sites. This involved counting the number of neighbouring atoms to the reaction active site rather than using the more conventional Sabatier principle to predict the activity of the catalyst based on the strength of the adsorption and desorption of key intermediates. Benedetti et al. [63] utilised the "coordination–activity plot" to design their nanozymes to structurally mimic the natural enzyme using a bimetallic alloy with etched nano-channels and demonstrated an improvement in the electrocatalytic performance of such a nanozyme.

In the following sections, the environmental applications of nanozymes will be discussed in greater detail. In particular, the administration of nanozymes for pollutant detection and degradation are intricate functions of the physicochemical properties of different types of nanozymes. Therein, (i) the ability to structurally manipulate the nanoparticles with atomic precision which affords specificity towards the detection of targeted pollutants, (ii) the surface engineering, such as the coatings applied, to improve the sensitivity of the detection and efficiency of biodegradation, and (iii) the transduction methods applied to enable the detection and biodegradation of pollutants are reviewed.

2. Detection of Metal Ions

2.1. Introduction

The persistence of toxic heavy metals in the environment is an important concern that can lead to adverse effects on human health. Heavy metals occur naturally in water, food and soil but also arise from industrial activities such as mining, smelting, electroplating, wood preserving and leather tanning [78]. In comparison to organic contaminants, inorganic ions are not biodegradable and tend to accumulate in the environment and organisms and enter the food supply chain. Effects from heavy metal exposure include nausea, vomiting, diarrhoea, abdominal pain, kidney dysfunction and, in extreme cases, death [78]. Therefore, there is a critical need for highly sensitive and selective sensors for monitoring metal ions. The Australian Drinking Water Guidelines for selected metals are listed in Table 2 [78].

Analytical instrumentation such as atomic absorption spectroscopy (AAS), inductively coupled plasma–optical emission spectroscopy (ICP-OES) and inductively coupled plasma–mass spectrometry (ICP-MS) [79] can address the needs of highly accurate and sensitive metal ion detection but require sample processing and analysis back in the laboratory as well as operation by trained personnel. Due to the high cost of this instrumentation, it are not easily accessible. Even where access to analytical laboratory testing facilities is available, the cost for analysis of a single sample is of the order of AUD 50–100 (USD 35–70) with a turnaround time of 2–5 days. This inherently limits the frequency of sampling as well as the number of collection points, which, in some instances, reduces the ability to rapidly identify the pollutant and its source.

Table 2. Australian Drinking Water Guidelines for Selected Metals [78].

Metal	Guideline Value (mg/L)	Potential Source
Arsenic	0.01	From natural sources and mining/industrial/agricultural wastes.
Cadmium	0.002	Indicates industrial or agricultural contamination; from impurities in galvanised (zinc) fittings, solders and brasses.
Chromium (VI)	0.05	From industrial/agricultural contamination of raw water or corrosion of materials in distribution system/plumbing.
Copper	2	From corrosion of pipes/fittings by salt, low-pH water.
Lead	0.01	Occurs in water via dissolution from natural sources or household plumbing containing lead (e.g., pipes, solder)
Mercury	0.001	From industrial emissions/spills. Very low concentrations occur naturally.
Silver	0.1	Concentrations are generally very low. Silver and silver salts occasionally used for disinfection.

Ideally, sensors which are portable and amenable for in-field sensing can reduce analysis times and provide real-time information as to the health state of the environment. Solid-state sensors that rely on colorimetric or electrochemical transduction can address these challenges as they are more compact and simpler to operate than lab-based analytical instrumentation. These sensors will ultimately lead to cost savings, as timely reporting allows rectifying actions to be taken before conditions worsen. Moreover, humans, animals and aquatic life will be prolonged by reducing the risk of exposure to these toxic elements.

Most colorimetric [80,81] or electrochemical transduction techniques [82,83] rely on a chemically sensitive material to interact with the metal ion of interest via coordination, adsorption or precipitation to induce a change in output signal. Detection of metal ions using metal nanoparticles generally involves the aggregation of nanoparticles to induce a colour change [84]. The aggregation of nanoparticles is facilitated by surface ligands which bind to the metal ion of interest and crosslink the nanoparticles, resulting in a shift in the surface plasmon resonance. Ionophores and porphyrins are also ligands that have been used to coordinate metal ions with high selectivity [85] and can be tethered directly to a substrate or immobilised within a polymer matrix. Not surprisingly, oligopeptides, DNA and enzymes [82,86–88] have also been studied for metal chelation since, in nature, metal ions play a key role in their interaction with living organisms. Metal ions can be highly toxic to enzymes since the thiol or methylthiol groups of amino acids near the active centre of enzymes can be blocked by heavy metals and inhibit their function [89].

In the past decade, several research groups have explored the use of nanozymes for the detection of metal ions [90–93]. Similar to natural enzymes, the peroxidase, oxidase or catalase-like activity of nanozymes can be inhibited or enhanced by the presence of metal ions. Many studies exploit the intrinsic peroxidase-like activity of nanozymes and are mainly type II metal nanoparticles [94–101]. These metal nanoparticles form an amalgam upon interaction with a metal ion or agglomerate through binding of the metal ion with the surface ligands of nanoparticles, changing the enzyme-like activity. The peroxidase-like activity of type I nanozymes such as cysteine-stabilised Fe_3O_4 magnetic nanoparticles (MNPs) [102], chitosan-functionalised $MoSe_2$ [103] and Co_9S_8 [104] can also be triggered by metal ions. Nanocomposite type III nanozymes have also been explored through the synergistic combination of metal nanoparticles, graphene-based materials or MOFs [56,105–109]. These nanocomposites can serve to enhance the catalytic activity or are dual-purpose nanozymes for the adsorption and detection of the metal ion of interest. Studies exploiting the oxidase [110,111] or catalase-like [112,113] activity of nanozymes for metal ion detection are limited. Whilst most peroxidase-like nanozymes exhibit high metal ion selectivity, the demonstrated examples of catalase-like nanozymes were cross-selective [112,113] with the ability to detect multiple metal ions under various conditions. The proceeding sections will review the various metal ions and their detection using nanozymes.

2.2. Mercury Detection

Mercury is a highly toxic heavy metal which naturally occurs in water, soil and even food. Due to its wide range of adverse effects on human health, including tremors, mental disturbances and gingivitis [78], many efforts have been focused on developing highly sensitive and selective Hg^{2+} sensors.

Long et al. [101] first reported a colorimetric sensor for Hg^{2+} based on the peroxidase-like enhancement of gold nanoparticles. TMB, a chromogenic substrate, can be oxidised by H_2O_2 in the presence of citrate-capped gold nanoparticles, resulting in a visual colour change from colourless to light blue. Remarkably, the addition of trace Hg^{2+} to the citrate-capped gold nanoparticles dramatically enhanced the oxidation of TMB, with the colour intensity proportional to the Hg^{2+} concentration. Compared to other metal ions at 25 times higher concentration, including K^+, Na^+, NH_4^+, Ag^+, Ba^{2+}, Mg^{2+}, Co^{2+}, Ni^{2+}, Zn^{2+}, Cu^{2+}, Mn^{2+}, Pb^{2+}, Cd^{2+}, Al^{3+}, Cr^{3+} and Fe^{3+}, the extent of oxidation of TMB was unaltered. The enhancement of peroxidase-like activity due to Hg^{2+} was attributed to two processes. In the first step, Hg^{2+} was reduced to Hg^0 by sodium citrate, which also acted as a stabilising agent for the nanoparticles. Subsequently, the reduced mercury dispersed across the gold nanoparticle surface, forming a Hg-Au amalgam, and improved the peroxidase-like activity of the nanoparticles. The altered nanoparticle surface enhanced the formation of •OH radicals on the surface due to bond breakage of H_2O_2. Similar phenomena were also observed with MoS_2-Au [108], Cu@Au nanoparticles [94] and Au/Fe_3O_4/graphene oxide [109]. In the latter case, the catalytic activity of Au/Fe_3O_4/graphene oxide was improved in the presence of Hg^{2+}, resulting in ultrasensitive detection of Hg^{2+} down to 0.15 nM [109]. Furthermore, mercury was able to be removed (>99% efficiency) from the surface of Au/Fe_3O_4/graphene oxide by the application of an external magnetic field, allowing reuse up to 15 times. In an approach towards rapid, on-site testing [114], Han et al. have developed a paper-based analytical device for Hg^{2+} detection based on citrate-stabilised gold nanoparticles (see Figure 7a). The device was designed so that multiple samples could be analysed simultaneously by having separate zones for reagent loading, absorbents and detection. By varying the number of drops of the test sample, colorimetric signal amplification due to the catalytic reaction of TMB and H_2O_2 could be realised.

Whilst the peroxidase-like activity of gold nanoparticles was stimulated by Hg^{2+}, the application of platinum nanoparticles resulted in enzyme inhibition. Similar to gold, the mechanism for a change in catalytic activity of platinum-based nanoparticles was a result of Hg-Pt amalgam formation at the Pt surface, affecting the electronic structure and active Pt^0 proportion. However, Hg^{2+} can interact directly with Pt^0 via metallophilic interactions due to their matched d^{10} configuration, resulting in catalytic inhibition [91]. Li et al. [95] were the first to demonstrate in 2015 that the peroxidase-like activity of Pt nanoparticles stabilised by bovine serum albumin (BSA) could be inhibited through interactions between Hg^{2+} and Pt^0, allowing the detection of Hg^{2+} down to 7.2 nM in 20 min without significant interference from other metal ions. Similarly, Zhou et al. [96] have used citrate-stabilised

Pt nanoparticles for the detection of Hg^{2+} ions. The OPD colorimetric signal was inhibited through the reduction of Hg^{2+} by citrate, resulting in a Pt-Hg amalgam (see Figure 7b). To enable screening in-field in various water bodies and public drinking water distribution systems, Kora and Rastogi [115] have developed a green approach for the synthesis of platinum nanoparticles using chloroplatinic acid and a nontoxic, biodegradable plant exudate gum as a reducing and stabilising agent. These nanoparticles (with an average size of 4.4 nm) showed excellent tolerance to temperature and pH changes and exhibited peroxidase-like activity.

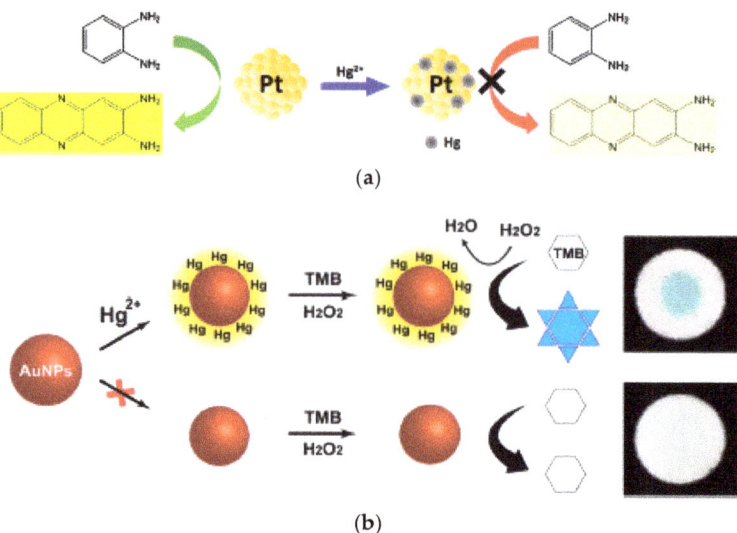

Figure 7. (**a**) Detection of Hg^{2+} ions by inhibition of OPD activity by formation of a Pt-Hg amalgam. (Reprinted with permission from [96]. Copyright © 2017, Elsevier B.V. All rights reserved.) (**b**) Detection of Hg^{2+} by enhancement of TMB activity by formation of an Au-Hg amalgam. (Reprinted from [114]. Copyright © The Authors. Distributed under a Creative Commons BY (CC BY) license).

Although the majority of Pt-based nanomaterials exhibit peroxidase-like activity, the oxidase-like activity has also been demonstrated, which is more ideal for environmental detection since unstable H_2O_2 is not required. Doping of Pt nanostructures with Se showed oxidase-like activity [110] due to an acceleration of electron transport and the anchoring of Se to Pt nanoparticles. The presence of Hg^{2+} inhibited the catalytic activity, enabling a facile colorimetric assay for Hg^{2+} down to 70 nM.

The enzyme-mimicking properties of nanomaterials can also be enhanced or inhibited through surface modification. Iron oxide nanoparticles are known to exhibit peroxidase-like activity, but their activity can be inhibited through blockage of the active Fe sites [6]. Cysteine-modified Fe_3O_4 nanoparticles [102] exhibited almost no colour change in the presence of TMB and H_2O_2 under pH 4.0 conditions. However, the presence of Hg^{2+} triggered the peroxidase-like activity of Fe_3O_4 due to the formation of a cysteine–Hg^{2+}–cysteine complex (Figure 8). Excellent sensitivity was achieved, with a detection limit of 5.9 nM. Detection of Hg^{2+} was also enhanced using gold nanoparticles modified with chitosan [97]. Hg^{2+} can react with chitosan via the NH_2 groups. By using TMB as a chromogenic substrate, and with 10-min incubation at 50 °C in pH 4.2 acetate buffer, the absorption peak at 652 nm was dramatically improved due to Hg^{2+}. Selenium nanoparticles have also been shown to catalyse the oxidation of TMB by surface modification with chitosan. Compared to BSA and sodium alginate Se nanoparticles, the chitosan Se nanoparticles had the highest activity, which was attributed to the involvement of reactive oxygen species that may react more favourably with chitosan [111]. The oxidase-like activity of chitosan-stabilised Se nanoparticles could be inhibited by Hg^{2+} due to the extremely high affinity of mercury for selenium. High selectivity for Hg^{2+} (5 µM) was afforded when evaluated under the same conditions as Cd^{2+}, Pb^{2+}, Cu^{2+}, Na^+, K^+, Ca^{2+}, Mg^{2+}, Zn^{2+}, Mn^{2+}, Al^{3+} and Fe^{3+} (100 µM). Similarly, Huang et al. [103] have demonstrated that chitosan-functionalised molybdenum (IV) selenide exhibited peroxidase and oxidase-like activities. The chitosan surface functionality enabled the reduction of Hg^{2+} ions to Hg^0 and further enhanced the enzyme activity. The selectivity of the nanozyme to Hg^{2+} (2 µM) was well-demonstrated over other ions at 10 µM.

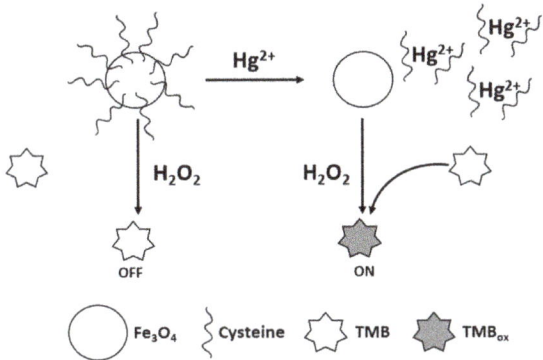

Figure 8. Principle for sensing of Hg^{2+} based on Hg^{2+}-triggered peroxidase-mimicking activity of Cys-Fe_3O_4 nanoparticles. (Adapted with permission from [102]. Copyright © 2018, Elsevier B.V. All rights reserved.).

The use of nanocomposites has also been gaining popularity as enzyme mimics. Combining metal nanoparticles with support materials such as graphene and MOFs is highly attractive as they may result in higher activities owing to synergistic interactions [56,105,108,109]. Gold nanoparticles dispersed onto a porous amino-functionalised titanium-based MOF exhibited peroxidase-like behaviour [105] that was higher than the individual gold nanoparticles and titanium-based MOF. Furthermore, kinetic analysis of the nanocomposite revealed a lower Michaelis constant value, K_m (i.e., a higher affinity for the substrate), than HRP. The presence of cysteine inhibited the peroxidase-like behaviour but could be reactivated by Hg^{2+} due to the strong affinity of Hg^{2+} for the thiol group of cysteine. The detection of Hg^{2+} was interfered with by Cu^{2+} and Ag^+ but could be partially masked with ethylenediaminetetraacetic acid (EDTA) and Cl^-. MOFs have also been decorated with Pt nanoparticles [56]. Combining the high peroxidase-like activity of Pt nanoparticles with the high porosity and surface area of MOFs provided a rapid technique for the measurement and removal of Hg^{2+} ions. Hg^{2+} ions inhibited the peroxidase-like activity, with high sensitivity and a Hg^{2+} detection limit of 0.35 nM, without significant interference from coexisting metal ions. Furthermore, by using 5 mg of the nanocomposite, effective removal (>99%) of Hg^{2+} (up to 50 mg/L) could be demonstrated after 12-h incubation. It is envisaged that advances in nanomaterials would result in the rise of more dual-purpose materials that can be used for both sensing and adsorption of metal ions.

2.3. Lead Detection

Lead is a cumulative poison that can have a major detrimental effect on the central nervous system [78]. Considering the potential health effects, the permissible Pb(II) level for Australian drinking water is set at 10 µg/L [78]. Thus, it is imperative to provide simple methods for monitoring Pb^{2+} in a variety of samples (environmental and food). Natural enzymes, such as DNAzymes [116], have high catalytic activity for Pb^{2+} and have been employed as biosensors for Pb^{2+}. In recent years, enzyme-mimicking nanomaterials for the detection of lead ions have gained popularity, involving enhancement and inhibition strategies. Gold nanoclusters modified with glutathione exhibit weak peroxidase-like activity [98] but the presence of Pb^{2+} ions can induce the aggregation of the gold nanoclusters due to binding with glutathione (Figure 9). The catalytic activity of TMB could be increased ten-fold upon aggregation, with a Pb^{2+} detection limit of 2 µM. Other bivalent metal ions (Ca^{2+}, Cd^{2+} and Zn^{2+}) capable of forming complexes with glutathione interfered with the detection of Pb^{2+}, but at much higher concentrations. Au@Pt nanoparticles [117] have also been used for the determination of Pb^{2+} ions through inhibition of the peroxidase-like activity. In the presence of sodium thiosulfate, Pb^{2+} ions accelerated the leaching of gold, which induced slight aggregation of the gold nanoparticles from a hydrate size of 35.5 to 75.0 nm. The aggregated nanoparticles exhibited weakened peroxide activity and could be correlated to Pb^{2+} concentration without interference from other metal ions. A detection limit of 3.0 nM Pb^{2+} was achieved. In another example, layered WS_2 nanosheets [118] also showed peroxidase-like activity which could be inhibited by Pb^{2+} ions. The high selectivity towards Pb^{2+} was attributed to the layered structure of the nanomaterial, which had a stronger adsorption capacity for Pb^{2+} than other metal ions. The nanozyme also offered high sensitivity (detection limit of 4 µg/L), which provided an easy method to distinguish whether the permissible Pb^{2+} level was exceeded even with the naked eye.

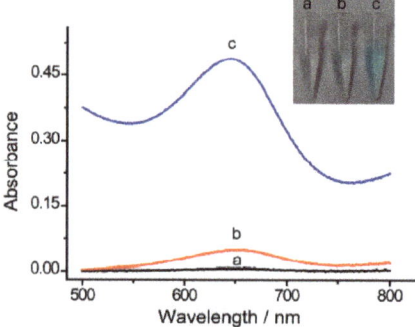

Figure 9. Absorption spectra and image (inset) of 500 µM TMB in the presence of (**a**) 10 mM H_2O_2, (**b**) 10 mM H_2O_2 and 5 µg/mL^{-1} Au nanoclusters, and (**c**) 10 mM H_2O_2, 5 µg/mL^{-1} Au nanoclusters and 250 µM Pb^{2+}. (Reprinted with permission from [98]. Copyright © 2017, Royal Society of Chemistry.).

2.4. Silver Detection

Silver is widely used in the electrical, photography and pharmaceutical industries and has resulted in increased levels of silver in the environment. Bioaccumulation of silver in the human body can lead to argyria, which results in bluish-grey metallic discolouration of the skin, hair, mucous membranes, mouth and eye [78]. Thus, it is vital to monitor the presence of silver in environmental and biological samples. Nanozymes for Ag^+ detection based on catalytic enhancement [106] and inhibition [99,100] have both been demonstrated. To more closely mimic natural enzymes, Zhang et al. [99] have modified Pd nanozymes with histidine. Amino acids are attractive surface modifiers for nanoparticles due to their biocompatibility and metal chelation ability. Histidine-modified Pd nanoparticles prepared by using histidine as a stabiliser and $NaBH_4$ as a reducing agent offered enhanced TMB colour formation (peroxidase-like activity) compared to bare Pd particles, which were formed in the absence of histidine. In the presence of Ag^+, specific binding to histidine resulted in the Pd nanoparticles being exposed, and the peroxidase-like activity was suppressed. A limit of detection down to 4.7 nM was achievable. No significant suppression of the TMB colour was observed in the presence of Hg^{2+}, K^+, Na^+, Mg^{2+}, Ca^{2+}, Mn^{2+}, Cd^{2+}, Ni^{2+}, Cu^{2+}, Au^{3+} or Fe^{3+} ions at the same concentration as Ag^+ (Figure 10). In another example of peroxidase-like inhibition, Au clusters stabilised by BSA were developed by Chang et al. [100]. The introduction of Ag^+ as low as 0.204 µM selectively reacted with Au^0 through redox reaction, which suppressed the oxidation of TMB by the BSA-stabilised gold clusters.

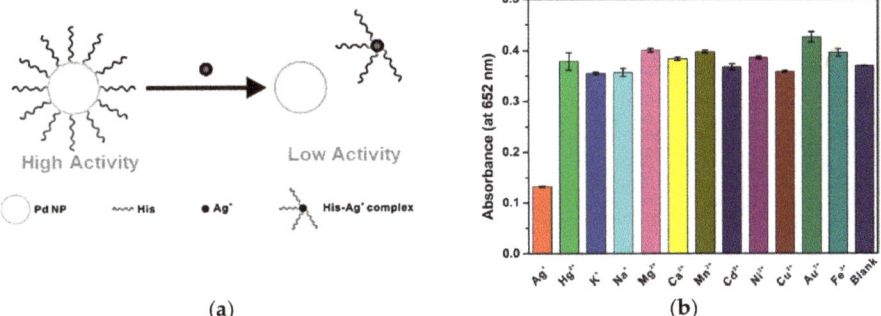

Figure 10. (**a**) Mechanism for the peroxidase-like activity of histidine-modified Pd nanoparticles and its effect in the presence of Ag^+. (**b**) Selectivity of the Pd nanoparticles for Ag^+ compared to other metal ions as demonstrated by a decrease in oxidised TMB colour formation. (Reprinted with permission from [99]. Copyright © 2018, Elsevier B.V. All rights reserved).

In order to shift away from the high cost of noble metal nanozymes, Li et al. [106] have developed zeolitic imidazolate frameworks-9/graphene oxide (ZIL-8-GO) as a peroxidase-mimicking nanozyme. The introduction of Ag^+ greatly enhanced the peroxidase-like activity and there was no significant interference from other metal ions even at 250 µM. Detection of Ag^+ was achieved visually by spotting filter paper with the reagents, offering the practicality of testing real samples. The Ag^+ detection limit using the naked eye was 0.1 µM, compared to 1.43 nM using UV–Vis spectrophotometry in a sample cuvette.

2.5. Copper Detection

Constant monitoring of Cu^{2+} in environmental and drinking water is necessary since an intake of excess Cu^{2+} can cause liver or kidney diseases [78]. Many colorimetric assays have been developed for Cu^{2+} sensing [119,120] but the use of nanozymes has been limited. Recently, urchin-like Co_9S_8 nanomaterials [104] with needle-like nanorods were found to be intrinsically catalytic due to their ability to facilitate electron transport as a variable-valence metal sulphide between TMB and H_2O_2. The addition of cysteine switched the Co_9S_8 sensor to its "off" state due to TMB radical restoration by the thiol functionality of cysteine. However, the presence of Cu^{2+} reactivated the sensor back to its "on" state, with the ability to achieve excellent selectivity and a Cu^{2+} detection limit of 0.09 µM. Glutathione can also influence the catalytic activity of nanozymes such as Pt nanoparticles and play an important role in Cu^{2+} detection [121]. Pt nanoparticles exhibit oxidase-like activity as they can effectively catalyse the oxidation of TMB by O_2. Glutathione inhibited the oxidase-like activity but its activity could be regained in the presence of Cu^{2+} due to oxidation of glutathione. Cu^{2+} detection was demonstrated in human serum but could also potentially be applied to environmental systems.

2.6. Chromium Detection

Elevated levels of chromium in the environment exist due to industrial activities such as the prevention of corrosion on metal surfaces [122]. Cr(VI) is highly toxic and can easily spread and bioaccumulate. Wang et al. [107] have developed an oxidase-like nanozyme that can detect Cr(VI) over a range of 0.03–5 µM with high selectivity. The presence of Cr(VI) and cerium oxide nanorod-templated MOFs boosted the oxidation of TMB substrate. The applicability of the nanozyme was demonstrated in spiked water samples, illustrating the potential for trace metal analysis in environmental waters.

2.7. Arsenic Detection

Arsenic poisoning stemming from long-term exposure to contaminated drinking water is a serious concern. In some parts of Australia [78], and in countries such as Argentina, Bangladesh, Chile, China, India, Mexico and the United States of America, concentrations of naturally occurring arsenic may exceed safe levels [123]. Despite the importance of arsenic detection, there are only a few examples of nanozymes being used for arsenic detection, all of which were published recently [124,125]. Cobalt oxyhydroxide (CoOOH) nanoflakes exhibit peroxidase-like activity [125] and can bind specifically to arsenate, As(V), via electrostatic attraction and As-O interactions. As(V) inhibited the oxidation of ABTS, resulting in a colorimetric detection limit of 3.72 ppb. By exploiting the redox conversion between ABTS and ABTSox, electrochemical detection of arsenate via chronoamperometry at a CoOOH-modified glassy carbon electrode resulted in even higher sensitivity, with a detection limit of 56.1 ng/L. Similarly, Zhong et al. [124] demonstrated an electrochemical and optical method for the determination of As(V) by using the peroxidase-like activity of iron oxyhydroxide (FeOOH) nanorods. The presence of As(V), as low as 0.1 ppb, inhibited the ability of the nanorods to catalyse the oxidation of ABTS substrate by hydrogen peroxide.

2.8. Detection of Multiple Metal Ions

Most of the reported nanozymes exhibited high selectivity, which is ideal for monitoring a specific metal ion. However, due to the coexistence of multiple metal ions in environmental waters, a simple test that can monitor a suite of metal ions is highly desirable. Depending on the nature of the nanozyme, some were sensitive to multiple metal ions. Metallothionein-stabilised copper nanoclusters displayed catalase-like activity [113] due to their ability to decompose H_2O_2 and inhibit the oxidation of TMB. Interestingly, Pb^{2+} and Hg^{2+} ions were able to induce the conversion of the catalase-like activity to peroxidase-like activity (Figure 11), with detection limits down to 142 and 43.8 nM, respectively. The assay displayed potential for simultaneously monitoring toxic Pb^{2+} and Hg^{2+} in environmental water samples.

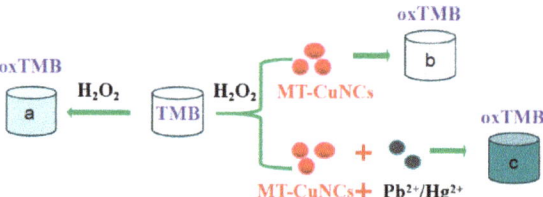

Figure 11. Oxidation of TMB by H_2O_2 (product a), catalase-like activity of metallothionein-stabilised copper nanoclusters (product b) and peroxidase-like activity of metallothionein-stabilised copper nanoclusters in the presence of Pb^{2+}/Hg^{2+} (product c). (Reprinted with permission from [113]. Copyright © 2019, Springer-Verlag GmbH Austria, part of Springer Nature).

Mercury and silver usually coexist in water, and peroxidase-like nanozymes have been developed that are sensitive to both metal ions. Surface modification of Au@Pt nanoparticles with sodium dodecyl sulfate effectively shielded most metal ions via complexation, except for Hg^{2+} and Ag^+ [126]. The detection limits for Hg^{2+} and Ag^+ were 3.5 and 2.0 nM, respectively. To discriminate Hg^{2+} and Ag^+, cystine was added to shield Hg^{2+} as a result of cystine–Hg^{2+} binding interaction (Figure 12). In a similar approach, Zhao et al. [127] have used EDTA to mask Hg^{2+}, providing a sensitive and selective means to detect both Hg^{2+} and Ag^+ using the peroxidase-mimicking ability of polyvinylpyrrolidone-coated platinum nanoparticles.

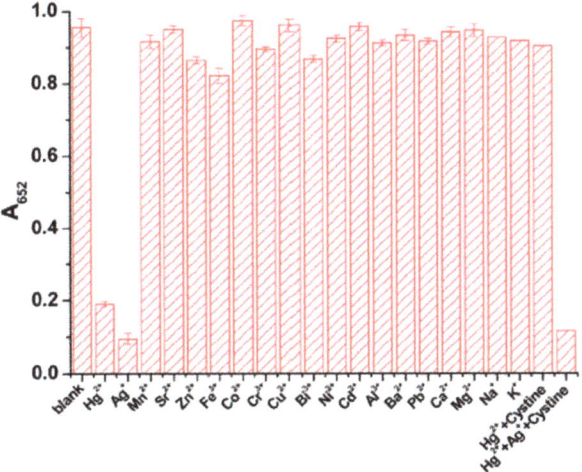

Figure 12. Selectivity of Au@Pt nanoparticles towards various metal ions in the presence of sodium dodecyl sulfate. (Reprinted with permission from [126]. Copyright © 2017, Royal Society of Chemistry).

Han et al. [112] have exploited the pH-sensitive catalase-like activity of Co_3O_4 MNPs to discriminate metal ions. The nanoparticles were insensitive to pH in the range of 6 to 11 but highly influenced by the nature of the metal ion (Ca^{2+}, Fe^{3+}, Hg^{2+} and Mn^{2+}). For instance, the catalase-like activity of Co_3O_4 MNPs was inhibited by Fe^{3+} at pH 7.0 but enhanced at pH 10.0. Conversely, the nanozyme was enhanced by Mn^{2+} at pH 7.0 but inhibited at pH 10.0. This could provide a powerful approach to discriminate metal ions (Figure 13).

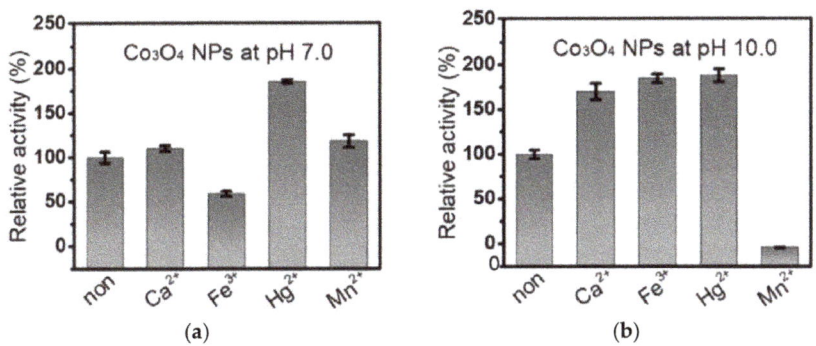

Figure 13. Influence of metal ions on the catalase-like activity of Co_3O_4 nanoparticles at (**a**) pH 7.0 and (**b**) pH 10.0. (Reprinted with permission from [112]. Copyright © 2018, John Wiley and Sons).

2.9. Summary

Nanozymes are highly promising materials for the determination of metal ions with high sensitivity and selectivity. The vast majority of the nanozymes were reliant on the peroxidase-like activity, which could be enhanced or inhibited in the presence of metal ions through amalgam formation or surface modifiers. However, those that are based on oxidase-like activity are more desirable since they do not rely on unstable H_2O_2 or its addition to environmental samples.

Nanozymes have demonstrated their applicability for in-field measurements where sample processing is conducted on-site. Most of the reported nanozymes are based on peroxidase-like activity and operate optimally under mildly acidic conditions (pH 4), requiring dilution in sodium acetate/acetic acid buffer solutions, with colour development ranging from a few minutes up to half an hour. Such sample processing can limit the practicality of in-field measurements, and although neutral pH operation is possible, this could be at the sacrifice of sensitivity. Nanozymes deposited on solid supports such as paper are attractive as they offer recyclability of the nanozyme, parallel sample measurements and signal amplification. Nowadays, advances in mobile phone technology have allowed the colour signal to be easily interpreted from built-in or custom-made applications, rather than relying on a UV–Vis spectrophotometer. However, accomplishing low detection limits with mobile phone technologies can be a challenge without extending the measurement time. Furthermore, approaches to more environmentally benign nanozymes or economically non-metal-based nanozymes are emerging. In particular, nanozymes that can serve multiple purposes, such as sensing and adsorption, are highly desirable in tackling environmental pollution.

3. Detection of Pesticides

3.1. Introduction

Pesticides have been extensively used in modern agriculture to control and eliminate pests by interfering with their metabolism, life cycle or behaviour. Their widespread use poses an enormous threat to human health and the ecosystem when released into the environment, food and water supplies. Exposure to pesticides is most likely through contact with crops or household products, via the skin, breathing or ingestion. Due to their high toxicity, environmental agencies have set maximum levels for pesticides in drinking and surface water [78].

Pesticides can be classified as insecticides, herbicides, fungicides or other types depending on their purpose, and they involve different classes of chemicals such as organophosphates, pyrethroids, carbamates, arsenic and nitrophenol derivatives [128]. Organophosphate pesticides (OPPs) are synthetic pesticides whose acute toxicity is associated with their ability to inhibit acetylcholinesterase (AChE) enzyme in the central nervous system, resulting in the accumulation of the neurotransmitter acetylcholine [129]. Symptoms of organophosphate exposure include headaches, nausea, diarrhoea and respiratory arrest.

Classical techniques for detecting pesticides include liquid or gas chromatography coupled with mass spectrometry (LC-MS, GC-MS) [130,131], which offer excellent sensitivity in the nanomolar range. However, they are not amenable to rapid on-field detection of pesticide residues and require operation by a highly trained technician. Enzyme activity inhibition methods are promising alternatives due to their ease in operation and rapid response. These assays rely on a change in enzyme activity upon exposure to pesticides. For instance, acetylcholinesterase and butyrylcholinesterase are irreversibly inhibited by OPPs, providing a means for indirect pesticide detection. A range of enzyme assays have been developed including colorimetric Ellman assays, electrochemical assays, fluorescence assays and chemiluminescence [132–135].

Enzymes also form the basis of bioremediation strategies to reduce the impact of pesticides in the environment by degrading and transforming pollutants into less toxic forms [136]. There are several types of enzymes involved in the detoxification of pesticides including oxidoreductases, hydrolases and lyases [136,137]. Oxidoreductase enzymes catalyse the transfer of electrons from one molecule to another and often require additional cofactors to act as electron donors, acceptors or both. Hydrolases are commonly involved in pesticide remediation by hydrolysing esters, peptide bonds, carbon–halide bonds, etc., and generally operate in the absence of redox cofactors, making them highly attractive for remediation. For example, alkaline phosphatase is a hydrolase enzyme responsible for removing a phosphate from organophosphate pesticides. Lyases are a smaller class of enzymes than oxidoreductases and lyases. They catalyse the cleavage of carbon–carbon bonds and carbon bonds with phosphorus, oxygen, nitrogen, halides and sulfate in the absence of redox cofactors and water. As bioremediation involves the use of microorganisms and their enzymes, several environmental parameters such as temperature, moisture content and pH can affect microorganism growth, which, in turn, can affect the rate of pollutant degradation.

3.2. Nanozymes for Pesticide Detection

Recent years have seen the emergence of nanozymes for the monitoring and degradation of pesticide residues on plants, crops, soil and water samples. Some of the strategies for pesticide detection are reliant on enzyme-like inhibition assays [138–140] or exploit the phosphatase-like activity for pesticide degradation [141–143]. Those that are based on the phosphatase-like activity are type I nanozymes involving nanoceria [141–143], whereas those based on the peroxidase-like activity are mainly type I Fe_3O_4 nanoparticles [140,144,145] or type II metal particles [138,146–148].

There are several studies which rely on inhibition of the oxidase or peroxidase-like activity of nanomaterials by pesticides. Nara and co-workers [138] have developed a sensitive and selective colorimetric assay for malathion detection using palladium–gold nanorods. The nanorods exhibit high peroxidase-like activity in the pH 2 to 6 range and better kinetic parameters than HRP. The peroxidase-like activity was quenched by the presence of malathion, with the OPD colour output being diminished. A low detection limit of 60 µg/L could be achieved, with no cross-reactivity from other analogous organophosphates or metal salts. Xia et al. [149] developed a colorimetric assay for pyrophosphate by exploiting the peroxidase-like activity of MoS_2 quantum dots. By aggregating the quantum dots via the addition of Fe^{3+}, the peroxidase-like activity could be enhanced. However, when pyrophosphate coexisted with Fe^{3+}, the enhancement effect diminished due to the strong coordination between pyrophosphate and Fe^{3+}. A detection limit of 1.82 µM pyrophosphate was able to be achieved. Kushwaha et al. [150] reported the oxidase-like activity of Ag_3PO_4 nanoparticles for the colorimetric detection of chlorpyrifos using TMB as a substrate. Ag_3PO_4 was able to oxidise chlorpyrifos to chlorpyrifos oxon and sulphide ions. The sulphide ions interacted with Ag_3PO_4 and inhibited the catalytic activity through a negative feedback loop. As a result, chlorpyrifos as low as 9.97 mg/L could be detected. In other work, Biswas et al. [146] have shown that the peroxidase-like activity of gold nanorods can be inhibited by malathion by interacting with the surface of the nanorods. They hypothesised that the positive charge of the nanorod surface coated with cetyltrimethylammonium bromide had a high affinity for the sulfanyl group of malathion. This interaction masked the enzymatic activity and enabled detection of malathion down to 1.78 mg/L. Other organophosphates such as chlorpyrifos and parathion lack a sulfanyl group; thus, the colorimetric assay for malathion was highly selective. Furthermore, potential interference from metal salts containing Zn^{2+}, Pb^{2+}, Co^{2+} or Mg^{2+} as sulfates or nitrates was less than 0.01%.

Highly specific strategies for pesticide detection using aptamers have been designed by Weerathunge et al. [147]. Firstly, they exploited the intrinsic peroxidase-like activity of tyrosine-capped silver nanoparticles. Subsequently, a chlorpyrifos-specific aptamer was incorporated onto the surface of the nanoparticles which switched the nanozyme to the "off" state. To realise sensing of chlorpyrifos, the nanozyme sensor was switched back to its "on" state due to aptamer desorption from the nanoparticle surface as a result of aptamer–chlorpyrifos binding (Figure 14). High specificity of the nanozyme sensor was afforded as the presence of other organophosphate pesticides did not lead to aptamer desorption and a detection limit as low as 11.3 mg/L was possible. Based on the same principle [148], tyrosine-capped gold nanoparticles as peroxidase-like nanozymes have also been used for the specific detection of acetamiprid. Compared to surface-enhanced Raman spectroscopy using silver dendrites [151], the detection of acetamiprid using nanozymes was five times more sensitive, with a detection limit of 0.1 mg/L.

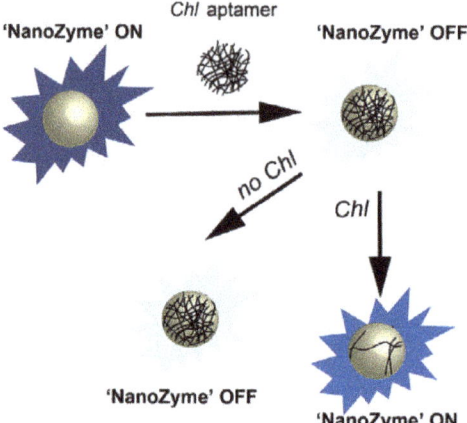

Figure 14. Working principle of tyrosine-capped silver nanoparticles used for the detection of chlorpyrifos (Chl). (Reprinted with permission from [147]. Copyright © 2019, Elsevier B.V. All rights reserved).

To target a suite of pesticides, nanozyme sensor arrays based on graphene oxide, nitrogen-doped graphene and sulphur-co-doped graphene with peroxidase-like activities have been developed (Figure 15) [139]. The interaction of the graphene materials with the aromatic pesticides lactofen, fluoroxypyr-meptyl, bensulfuron-methyl, fomesafen and diafenthiuron decreased their peroxidase-like activities. Molecular dynamics calculations confirmed that the enzyme-mimicking active sites (graphitic nitrogen in nitrogen-doped graphene, and carboxyl groups in graphene oxide) were blocked by the pesticides. Discrimination of the five pesticides (at 11 different concentrations) was demonstrated by using the three types of graphene prepared in a 96-well plate along with TMB and H_2O_2 in pH 4.0 sodium acetate buffer. By using linear discriminant analysis, the colorimetric response patterns were transformed into 2D canonical score plots which showed good clustering of the pesticides into five groups.

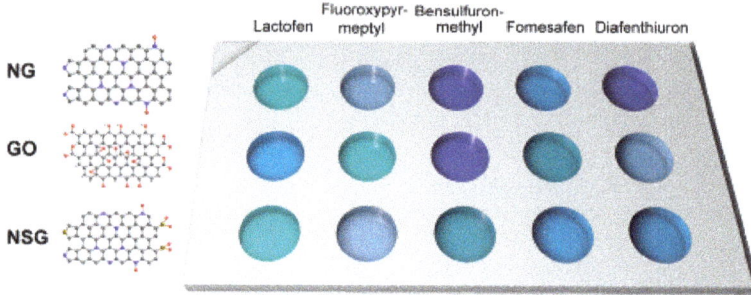

Figure 15. An array of peroxidase-mimicking graphene materials (graphene oxide (GO), nitrogen-doped graphene (NG) and sulphur-co-doped graphene (NSG)) used for the detection of five types of aromatic pesticides in the presence of TMB and H_2O_2. (Reprinted with permission from [139]. Copyright © 2020, American Chemical Society).

There have also been significant advances in using the peroxidase-like activity of Fe_3O_4 nanoparticles for the detection of pesticides. Guan et al. [144] have reported the chemiluminescent switching of Fe_3O_4 nanoparticles in the presence of pesticides. The nanoparticles catalyse the decomposition of dissolved oxygen to generate superoxide anions, enhancing the chemiluminescent intensity of luminol. The chemiluminescent signal could be quenched by the addition of ethanol as a radical scavenger; however, it was inhibited by a non-redox pesticide, ethoprophos. The strong surface coordinative reactions enabled the detection of ethoprophos down to 0.1 nM. Structurally analogous pesticides such as profenofos were also able to switch on the chemiluminescence, whereas the signal intensity from dylox and 2,4-dichlorophenoxyacetic acid was much lower. Therefore, the high specificity of the luminol-Fe_3O_4 system was attributed to the detection of organophosphorus esters with a P-S bond. Liang et al. [140] have also developed an assay for organophosphates based on Fe_3O_4 nanoparticles as a peroxidase mimic in combination with the enzymes AChE and choline oxidase (CHO). AChE and CHO catalyse the formation of H_2O_2 in the presence of acetylcholine, which can then be detected colorimetrically by Fe_3O_4 nanoparticles using TMB as a substrate. The reaction scheme is detailed in Equations (6)–(8). The presence of organophosphate compounds acephate and methyl-paraoxon as representative pesticides, and nerve agent Sarin, inhibited the activity of AChE and decreased the colour output. Compared to the traditional enzyme activity-based methods, the Fe_3O_4 peroxidase-like nanoparticles were more sensitive due to the catalytic activity of the nanoparticles, allowing concentrations as low as 1 nM Sarin, 10 nM methyl-paraoxon and 5 μM acephate to be detected. Boruah and Das [145] have successfully demonstrated that Fe_3O_4-TiO_2/reduced graphene oxide nanocomposites could be used as a colorimetric assay for atrazine as well as photocatalytic degradation of atrazine. The nanocomposite was highly effective towards the oxidation of TMB at a pH of 3. However, the absorbance intensity at 652 nm was diminished by atrazine, which was attributed to hydrogen bonding between the pesticide and TMB. The detection limit for atrazine was found to be 2.98 μg/L. The degradation of atrazine was monitored by UV–Vis spectroscopy at 221 nm and it was shown that over 99% degradation was achieved within 40 min. This dual-responsive material is highly promising as it is also magnetically separable and could be recycled up to ten times.

$$\text{acetylcholine} + H_2O \xrightarrow{\text{AChE}} \text{choline} \quad (6)$$

$$\text{choline} + O_2 \xrightarrow{\text{CHO}} H_2O_2 \quad (7)$$

$$H_2O_2 + \text{TMB} \xrightarrow{Fe_3O_4} \text{oxidised TMB} \quad (8)$$

Organophosphorus hydrolase is a useful phosphatase enzyme for detecting and degrading organophosphate pesticides with high specificity. As a degradation strategy, Wei et al. [141] have used nanoceria as a phosphatase mimic for the hydrolysis of organophosphate pesticides to *p*-nitrophenol using methyl-paraoxon as a representative compound. The hydrolysed product exhibited a bright yellow colour, which was analysed spectroscopically and with a smartphone. Under the optimal condition of pH 10, a detection limit of 0.42 μM was achieved. Dried plant samples were analysed by extracting the pesticide residues with ethyl acetate, evaporated to dryness and reconstituted in water. Potential interferents such as Na^+, K^+, Mg^{2+}, glucose, alanine, ascorbic acid, sodium acetate and tyrosine were evaluated. All substances except ascorbic acid demonstrated negligible interference. The same researchers [142] have also combined the remarkable phosphatase-mimicking activity of nanoceria with carbon dots for the fluorometric determination of pesticides. Carbon dots are attractive as fluorescent probes as they exhibit low toxicity compared to conventional semiconducting quantum dots. The hydrolysis of methyl-paraoxon to *p*-nitrophenol is yellow and largely overlaps with the excitation spectra of carbon dots. Thus, fluorescence quenching of the carbon dots was observed by the generated *p*-nitrophenol. The limit of detection for *p*-nitrophenol was calculated to be 0.376 μM. Electrochemical detection of the degradation product of methyl-paraoxon has also been demonstrated using nanoceria. Sun et al. [143] have exploited the bifunctionality of nanoceria as a phosphatase mimic to degrade methyl-paraoxon to *p*-nitrophenol, followed by electrochemical detection of *p*-nitrophenol at a nanoceria-modified glassy carbon electrode (Figure 16). The electrochemical method was highly sensitive, with a methyl-paraoxon detection limit of 0.06 μM.

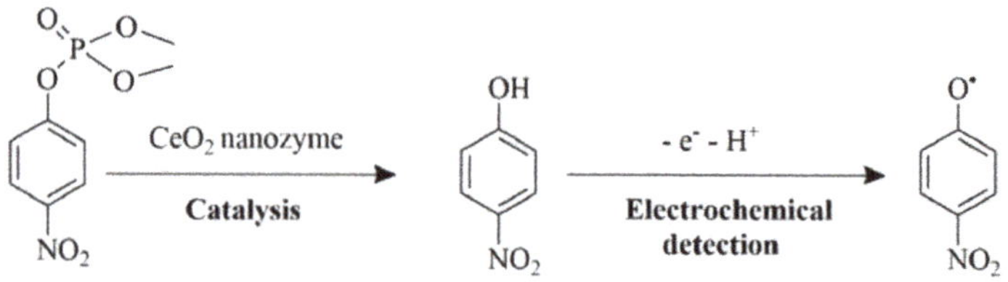

Figure 16. Schematic of nanoceria as a bifunctional material for catalysis and electrochemical detection of methyl-paraoxon. (Reprinted with permission from [143]. Copyright © The Authors. Distributed under a Creative Commons (CC BY-NC-ND 4.0) license).

3.3. Summary

With further development, colorimetric assays can be configured for rapid, low-cost, in-field detection of pesticides. However, the sensitivity towards trace pesticide residues remains a significant challenge as most nanozymes can only achieve sub-micromolar detection limits. Strategies to amplify the signal and improve specificity include aptamer surface modification and the use of molecularly imprinted polymers. Electrochemical and fluorometric methods have also been explored as alternate sensitive transduction strategies.

Although the detection of pesticides is important, their removal is also paramount since they can last several years in the environment before breaking down. Nanozymes have emerged with the capability to breakdown or degrade pesticides into more benign products and will be covered in more detail in Section 5. This section has covered the use of nanozymes for the detection of pesticides, the most prevalent type of persistent organic pollutant. Section 4 will discuss methods to detect other types of persistent organic pollutants.

4. Detection of Other Persistent Organic Pollutants

Persistent organic pollutants are compounds that can persist in the environment for extensive periods and be transported by wind and water or through the food chain. They include organophosphorus compounds, phenolic compounds, dyes and antibiotics. As discussed in the previous section, organophosphorus pesticides are the most widely encountered persistent organic pollutant. Phenolic compounds such as chlorophenols and bisphenols are also broadly used as pesticides, including as wood preservatives and disinfectants [152], and are commonly detected in ground water and soil. They are a major cause for health concern as they can be carcinogenic, neurotoxic, affect the reproductive system and disrupt the endocrine system. For the detection of phenol [153], Barrios-Estrada used Pt_3Au_1 nanoparticles decorated with few-layer MoS_2 nanosheets as peroxidase mimics. The oxidative coupling of phenol with 4-aminoantipyine in the presence of H_2O_2 resulted in the formation of a pink colour, with an optimum pH of 8.0–9.0. Nanozymes have also been shown to be highly promising candidates for the degradation of phenols [154–158] and will be discussed in Section 5. The efficiency in the removal and degradation of textile dyes will be covered in Section 5. Many organic dyes are toxic and not easily degraded in wastewater treatment plants [159]. These dyes serve as chromogenic substrates for the nanozymes, with the colour diminishing over time [160,161]. Antibiotics are classified as emerging pseudo-persistent organic pollutants as they are resistant to biodegradation due to their antimicrobial nature [162]. They are used in human and veterinary medicine and are mainly released into the environment through excretion [163]. Antibiotics are also used in agriculture, resulting in their presence in animal-derived food products, and can lead to serious side effects such as allergic reactions, hearing loss and kidney damage [164]. Zhao et al. [164] have developed an aptamer-modified gold nanoparticle sensor for the colorimetric detection of streptomycin. The gold nanoparticles exhibited peroxidase-like activity, which was diminished by coverage with streptomycin-specific aptamers. The presence of streptomycin as low as 86 nM resulted in re-establishment of the colorimetric signal from ABTS due to the formation of a streptomycin–aptamer complex. The sensor was highly specific, with little interference from tetracycline, oxytetracycline, carbamazepine, penicillin, amgoxicillin and diclofenac. Despite the gold nanoparticles appearing red in colour, a greyish-green colour was well observed from the streptomycin sample (Figure 17). Sharma, Bansal and co-workers [165] have similarly exploited the high specificity of ssDNA aptamers coupled with the intrinsic peroxidase-like activity of gold nanoparticles for the detection of kanamycin. A rapid visual readout was possible within 3–8 min, with a detection limit of 1.49 nM.

Figure 17. Selectivity of aptamer-modified gold nanoparticles for streptomycin. From left to right: streptomycin, amoxicillin, tetracycline, oxytetracycline, carbamazepine, diclofenac and penicillin added to aptamer-modified gold nanoparticles, ABTS and H_2O_2. (Reprinted with permission from [164]. Copyright © The Authors. Distributed under a Creative Commons (CC BY-NC) license).

5. Environmental Remediation with Nanozymes

5.1. Introduction

Human activities have resulted in significant contamination of our environment [3,166,167]. Many toxic pollutants, including heavy metals, dyes, other organic and inorganic substances, are generated during the production of items for our consumption and also during their use. These include pesticides, pharmaceuticals and clothes. Due to a lack of regulation and awareness, many toxic compounds have been discharged into the environment, particularly aquatic environments (lakes, rivers and oceans) without a second thought. These pollutants have caused significant damages to our ecological systems, including humans, plants, animals and microbes. They have been established to be the cause of toxic and carcinogenic effects on humans through contamination of drinking water and foods [167]. Significant efforts have been devoted to the remediation work to remove these pollutants and/or degrade them to less harmful products [92,166,168].

Methods for the removal of contaminants from aquatic environments include adsorption, membrane filtration, distillation, oxidation, biocatalytic and photocatalytic degradation [166,168]. Biocatalytic methods (enzymatic and microbial) have been investigated as effective means for the degradation of organic pollutants. The advantages of these biocatalytic methods are that they can be operated under mild and natural environments. Additionally, microbes and enzymes themselves could decompose into benign compounds when their mission is completed, thus eliminating unwanted environmental pollution from the treatment agent itself [136,137]. However, microbes and enzymes can only be functional and effective in narrow thermal and pH windows. The cost of producing enzymes, their lack of recyclability and the fact that biodegradation can be quite slow have hindered their large-scale, widespread application in environmental pollutant remediation [92,166].

Efforts to overcome the limitations of enzymatic methods have driven investigations into using nanozymes as remediating agents for environmental pollutants. Nanozymes have been shown to demonstrate catalytic properties, e.g., peroxidase- and oxidase-like, which are utilised in the degradation of pollutants by natural enzymes. Nanozymes could overcome enzyme limitations in terms of cost of production, recyclability, higher rate of reactions and wider operational windows (pH and temperature) [3,92,168]. The previous section has shown that nanozymes are effective for the detection of heavy metals and organic pollutants in the environment. Nanozymes could also be used to degrade environmental pollutants. In this section, applications and future prospects of nanozymes in environmental pollutant remediation will be discussed, focusing on the degradation of persistent organic pollutants.

Persistent organic pollutants include phenolic compounds, pesticides, dyes and organophosphorus compounds [169]. Phenolic compounds, particularly chlorophenols, have been widely used in pesticides, dyes and other synthetic compounds [156,168,170]. Chlorophenol pollutants can cause serious problems for the environment as they are highly toxic and resistant to chemical and biological degradation in the environment. These compounds are classified as top priority pollutants by the US EPA (Environmental Protection Agency) and other environmental regulators around the world. Current methods for phenol removal include solvent extraction, physical adsorption, pervaporation, wet air oxidation, ozonolysis, wet peroxide oxidation, electrochemical oxidation, photocatalytic oxidation, supercritical water gasification, electrical discharge degradation and bio-degradation [156]. Nanozyme-mediated degradation of phenolic compounds is a highly promising method with significant advantages over current methods [168].

There are many types of textile dyes currently in use (including phenolic dyes), and many of these are recalcitrant and toxic compounds that are not easily degraded in wastewater treatment plants [159]. On a global scale, approximately 700,000 tonnes (in 2005) of textile dyes are consumed yearly [171]. Textile manufacturing activities are currently concentrated in developing countries, where the treatment of organic dyes will need to be low-cost to be more widely applicable. Here, nanozymes could have significant advantages due to the low cost and high recyclability of certain nanozyme materials such as magnetic Fe_3O_4.

Organophosphorus compounds (OPPs, chemical warfare nerve agents and flame retardants), particularly OPPs, widely exist in the environment [136,172,173]. The increased use of highly toxic OPPs has made these compounds major contaminants in water, fruit and vegetables. OPPs have highly detrimental effects on human health and hence a method to remove them from the environment is invaluable [141].

Herein, the application of nanozymes in organic pollutant remediation will be discussed according to the nanozyme type classifications as proposed in the Introduction of this review.

5.2. Type I Nanozymes: Active Metal Centre (of Metalloenzyme) Mimics

Table 3 lists recent studies of type I nanozymes for the remediation of environmental pollutants. Since the report by Gao and co-workers [6] that Fe_3O_4 MNPs possessed enzymatic-like activity similar to naturally occurring peroxidases, they have been widely used for the oxidation of organic substrates as a detection method in the treatment of wastewater. Iron oxide nanoparticles, especially Fe_3O_4 MNPs, have been most widely investigated for the degradation of various environmental pollutants due to their peroxidase-like activity and the fact that Fe_3O_4 MNPs could be conveniently prepared from cheap and abundant precursors [168].

As can be seen in Table 3, the overwhelming enzyme-like catalytic activity marshalled by type I nanozymes in the degradation of phenol compounds and dyes is the peroxidase-like activity. Of the 17 examples listed above, only two examples where MnO_2 nanoparticles and Cu complex displayed laccase-like activity are non-peroxidase-like examples [172,173]. Laccases promote the oxidation of phenolic compounds with the reduction of oxygen to water. Fe_3O_4 was the most common material of the metal oxides investigated for environmental pollutant degradation.

The first example of using ferromagnetic nanoparticles to facilitate the decomposition of phenols was reported in 2008 [154]. Phenols were removed from wastewater by the peroxidase activity of the Fe_3O_4 nanoparticles. The hydroxyl radical formed in the reaction between Fe_3O_4 and hydrogen peroxide catalytically degraded phenols. More than 80% of phenols were removed at 16 °C and pH 3 and the nanozyme could be reused several times. Fe_3O_4 MNPs (5.7 nm in size) could also effectively degrade and mineralise 2,4-dichlorophenol. Fe_3O_4 MNPs were effective as heterogeneous sono-Fenton catalysts for the degradation of bisphenol A (BPA) in a reasonably wide pH range of 3–9. However, the rate of the degradation was still too slow for practical application [174]. Magnetic Fe_3O_4 nanoparticles at 30 nm in size were demonstrated to be highly effective for the degradation of 4-chlorophenol by Cheng et al. [75]. In the presence of hydrogen peroxide and Fe_3O_4, 4-chlorophenol was degraded to Cl^-, HCOOH (formic acid), CH_3COOH (acetic acid) and by-products. Adsorption of 4-chlorophenol onto the iron oxide particle surfaces was shown to be only minimal (around 10%) and an acidic pH of 5 was the optimal pH for the highest activity. These Fe$_3$O$_4$ nanoparticles could be reused with no decrease in reactivity. In fact, an increase in their reactivity for the degradation of 4-chlorophenol was observed. A plausible mechanism was proposed for the catalytic degradation of 4-chlorophenol (Figure 18). Fe$_3$O$_4$ converted H$_2$O$_2$ via its peroxidase-like activity, to form the highly active hydroxyl radical HO$^\bullet$, which, when reacted with 4-chlorophenol, triggered a chain of reactions leading to the formation of chloride ion, formic acid, acetic acid and other by-products.

Table 3. Examples of type I nanozymes in environmental pollutant remediation.

Nanozyme	Pollutant	Enzyme-Like Activity	Year	Ref.
Fe$_3$O$_4$ MNPs	Phenol	Peroxidase	2008	[154]
Fe$_3$O$_4$ MNPs	Rh B	Peroxidase	2010	[175]
Fe$_3$O$_4$ MNPs	Sulfamonomethoxine	Peroxidase	2011	[176]
Humic acid-Fe$_3$O$_4$ MNPs	Sulfathiazole	Peroxidase	2011	[177]
Fe$_3$O$_4$ MNPs	BPA	Peroxidase	2012	[174]
Fe$_3$O$_4$ MNPs	2,4-Dichlorophenol	Peroxidase	2012	[178]
CuO porous structures	Phenol	Peroxidase	2014	[155]
Fe$_3$O$_4$ MNPs	4-Chlorophenol	Peroxidase	2015	[170]
Fe$_3$O$_4$ nanorod bundles	Crystal violet	Peroxidase	2015	[160]
CuO nanoparticles	Phenol	Peroxidase	2015	[156]
Fe$_2$O$_3$·0.5H$_2$O (ferrihydrite) and hematite (Fe$_2$O$_3$)	Methylene blue	Peroxidase	2016	[161]
VO$_x$ nanoflakes	Rh B	Peroxidase	2016	[179]
MnO$_2$ nanomaterials	ABTS	Laccase	2017	[172]
MNPS@chitosan	Phenol	Peroxidase	2018	[157]
Fe$_3$O$_4$ nanorods	Rh B, methylene blue and methyl orange	Peroxidase	2019	[180]
Cysteine-histidine Cu	2,4-Dichlorophenol and other phenolic compounds	Laccase	2019	[173]
CeO$_2$ nanoparticles	Rh B, fluorescein, xylene cyanol FF, Brilliant Blue G-250, and Coomassie Brilliant Blue R-250	Peroxidase	2020	[181]

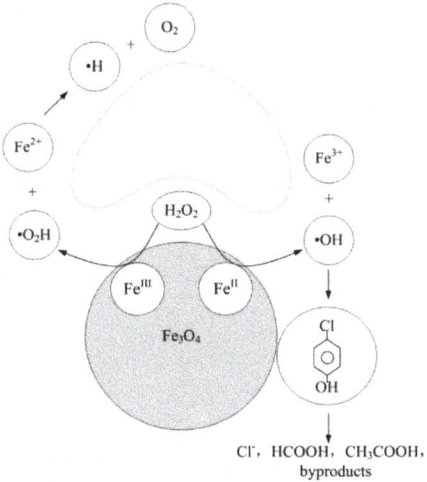

Figure 18. Illustration of the mechanism of catalytic oxidation of 4-chlorophenol by Fe$_3$O$_4$ nanoparticles. (Reprinted with permission from [75]. Copyright © 2019 WILEY-VCH Verlag GmbH & Co. KGaA, Weinheim, Germany).

Fe$_3$O$_4$ MNPs tend to aggregate and hence lead to the reduction of catalytic activity. To increase the catalytic ability and stability of Fe$_3$O$_4$ MNPs, Jiang and co-workers [157] synthesised ferromagnetic chitosan (MNP@chitosan, an alkali polysaccharide) nanozyme with particle sizes of around 12 nm. The MNP@chitosan was shown to be effective for the decomposition of phenol. The optimum pH for this catalytic activity was pH 4. The MNP@chitosan was also found to be recyclable; however, the degradation rate slowly decreased with each reuse cycle.

Fe$_3$O$_4$ MNPs or nanorods have also been successfully used in the degradation of Rhodamine B (Rh B) [175], crystal violet [160,180], methylene blue [161,180] and methyl orange [180] in the presence of hydrogen peroxide with peroxidase activity. Humic-acid-coated ferromagnetic nanoparticles (Fe$_3$O$_4$ 10–12 nm in size) were effective for the removal of sulfathiazole from aqueous media in the presence of H$_2$O$_2$. The catalytic efficiency was higher at lower pH (as low as pH 3.5) and higher temperatures, up to 60 °C [177].

Fe$_3$O$_4$ MNPs can activate persulfate anion S$_2$O$_8^{2-}$ to generate the powerful sulfate radical SO$_4^{\bullet -}$, which has high potential for the degradation of organic pollutants. The method was demonstrated to degrade sulfamonomethoxine (an antibiotic) in the presence of S$_2$O$_8^{2-}$ [176].

Ferric oxides (Fe$_2$O$_3$) such as 2-line ferrihydrite (2LFh, Fe$_2$O$_3 \cdot 0.5$H$_2$O, average size 5 nm) and hematite (Fe$_2$O$_3$, average size 100 nm) could also catalyse the degradation of methylene blue [161]. The more highly crystalline hematite was found to be more effective in degrading methylene blue than ferrihydrite. The author concluded that the ordered crystal planes of the hematite nanoparticles had a greater influence on the catalytic activity than the high surface area of 2LFh nanoparticles. A slightly basic pH of 8 was the optimal pH.

Other metal oxide nanoparticles have also shown peroxidase-like activity in the degradation of organic pollutants. Examples include CuO [155,156], CeO$_2$ [181], MnO$_2$ [172] and VO$_x$ [179].

CuO micro-/nanostructures with clean surfaces were prepared and demonstrated to have high peroxidase activity using TMB as the model substrate. These CuO structures could also degrade phenol, illustrating their promise as a reagent for wastewater treatment [155]. Feng and co-workers have shown cupric oxide nanoparticles to be highly efficient in the degradation of phenol, catechol, hydroquinone and other by-products [156]. Larger particles (30 nm) were found to be inefficient for phenol degradation. While no explanation was provided, the enhanced surface area to volume in smaller particles is likely to be a contributing factor. The optimal pH range was 3–7; above pH 7, the catalytic activity dropped off quite rapidly. In a highly acidic environment (pH 2), the phenol removal efficiency was very low as CuO particles dissolved to Cu^{2+} ions, which have low peroxidase activity.

Wang et al. [173] constructed a laccase-mimicking nanozyme by coordinating Cu$^+$/Cu^{2+} to cysteine-histidine dipeptide to form an inorganic polymer CH-Cu which possessed good catalytic efficiency for the degradation of chlorophenol and bisphenol in the absence of H$_2$O$_2$ (Figure 19). CH-Cu was also highly robust and could operate at pH 3–9, temperatures −20 to 90 °C, high salinity and could be stable for more than 3 weeks. The CH-Cu nanozyme could be reused multiple times.

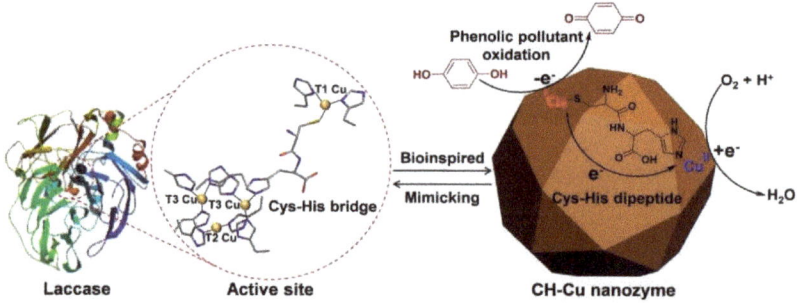

Figure 19. Cysteine-histidine Cu nanozyme for the oxidation of phenolic pollutant. (Reprinted with permission from [173]. Copyright © 2019, Elsevier B.V. All rights reserved).

Wang and co-workers showed that the degradation of various organic dyes such as Rh B, fluorescein and Brilliant Blue G-250 in the dark by CeO_2 could be enhanced using fluoride ions at low-pH conditions [107].

Manganese oxide, MnO_2, nanomaterials were demonstrated to have laccase-like catalytic reactivity and could catalyse the oxidation of pollutants in wastewater treatment in the absence of H_2O_2. G-MnO_2 was the most efficient catalyst for the degradation of ABTS and 17-β-estradiol (E2) amongst the different manganese oxide nanomaterials tested [172] (Figure 20).

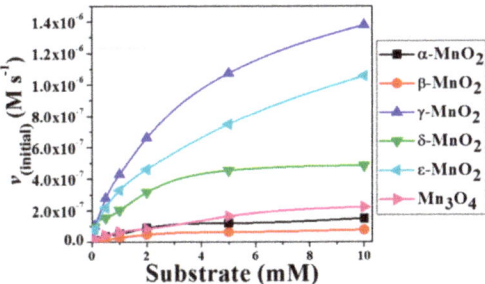

Figure 20. Degradation of ABTS with different manganese oxide materials. (Reprinted with permission from [172]. Copyright © The Authors. Distributed under a Creative Commons (CC BY) license).

Zeb et al. [179] reported that mixed-phase VO_x nanoflakes could be conveniently prepared in a single step and be a highly effective Fenton reagent, which could fully decompose Rh B within 60 s (Figure 21). The VO_x nanoflake also possessed peroxidase-like activity and could oxidise TMB efficiently, with V_{max} of approximately 27 times more than HRP. These results indicated that VO_x nanoflakes could be used for the effective degradation of environmental pollutants.

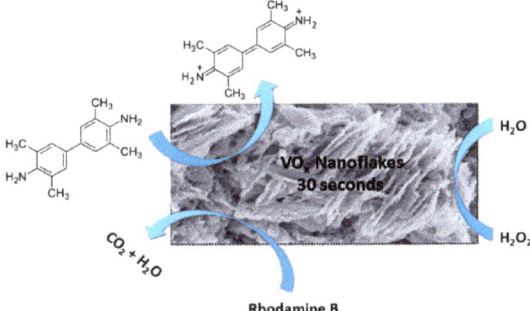

Figure 21. VO_x nanoflakes as efficient catalysts for the oxidation of TMB and decomposition of Rh B. (Adapted with permission from [179]. Copyright © 2016, American Chemical Society).

5.3. Type II Nanozymes: Functional Mimics

There are few examples of type II nanozymes that have been shown to degrade environmental organic pollutants (Table 4). Peroxidase-like activity was the dominant activity observed [182–184]; however, oxidase-like activity [185] was also observed.

Table 4. Examples of type II nanozymes in environmental pollutant remediation.

Nanozyme	Pollutant	Enzyme-Like Activity	Year	Ref.
Fe(0) nanoparticles	4-Chloro-3-methyl phenol	Peroxidase	2011	[183]
Carbon nanodots	Azo dyes (methyl red and methyl orange)	Peroxidase	2012	[182]
Cubic boron nitride	Rh B	Peroxidase	2016	[184]
Graphitic carbon nitride	TMB	Dual oxidase-peroxidase	2019	[185]

Xu et al. [183] investigated Fe(0) particles as heterogeneous Fenton-like catalysts for the removal of 4-chloro-3-methyl phenol. It was found that lower pH led to higher catalytic activity and the reaction was still quite efficient up to pH 6.1. A range of intermediates and products were observed by LC-MS, and the final products after 60 min of reaction included chloride ion, oxalic acid, acetic acid and formic acid. The catalytic activity decreased with time and the recyclability of the material was not demonstrated. It is worth noting that Fe(0) was oxidised to Fe(IV) (FeO^{2+}) during the catalytic reaction; hence, the Fe(0) particles are best considered to be a pre-catalyst in this work.

In a communication, Safavi and co-workers [182] reported the preparation of carbon nanodots using a microwave-assisted method in ionic liquids. These carbon nanodots could degrade azo dyes methyl red and methyl blue via their peroxidase-like activity.

Chen et al. [184] showed that cubic boron nitride possessed peroxidase-like activity and could oxidise TMB. The cubic boron nitride could be reused multiple times. The catalytic activity was highest at acidic pH 5 and at temperatures around 40–50 °C. Cubic boron nitride also catalysed the degradation of Rh B in the presence of H_2O_2.

Zhang and co-workers [185] reported modified graphitic carbon nitride as a metal-free nanozyme which has dual oxidase–peroxidase functions as a cascade photocatalyst for the oxidation of TMB. Whilst the work was not directed towards applications in environmental pollutant degradation, the fact that TMB (a substrate in the colorimetric test that has a similar structure to many organic dyes) is efficiently oxidised to its blue form is a strong indication that the graphitic carbon nitride is likely to be effective in dye removal. The bifunctional oxidase–peroxidase activities of the material would be highly advantageous in large-scale applications as there would be no need for hydrogen peroxide.

Noble metal nanozymes have been shown to be applicable for the detection of environmental pollutants and in medicinal diagnosis. They have not been widely applied in the degradation of environmental pollutants. The high cost of noble metals is likely to be a prohibitive factor in the application of noble metal nanozymes for environmental pollutant remediation. Consequently, there have not been many studies on using noble metal (Au, Pt etc.)-based nanozymes for environmental pollutant remediation.

5.4. Type III Nanozymes: Nanocomposites

Recently, nanocomposite materials including metal/metal oxides on carbon materials, MOFs and bimetallic alloys (core-shell) have been investigated extensively as agents for environmental remediation. As shown in Table 5, peroxidase-like activity is the predominant catalytic activity utilised by composite nanozymes in the degradation of organic pollutants. The recalcitrant organic pollutants investigated in these studies included phenolic compounds [48,158,186–189], dyes [159,190–196] and organophosphorus compounds [197,198].

Magnetic Fe_3O_4 on carbon materials [48,186–188,191,192,195,199–202] are the most studied class of composite nanozymes for environmental degradation. Carbon materials by themselves have been used in wastewater treatment for a long time. Composites of Fe_3O_4 with carbon materials are expected to deliver significant benefits such as recyclability and dual adsorptive–catalytic activities in pollutant degradation applications. The excellent review by Ribeiro et al. [202] is an important source of information for readers who have special interest in the application of hybrid magnetic carbon nanocomposites for the degradation of organic pollutants in water treatment.

Table 5. Examples of type III nanozymes in environmental pollutant remediation.

Nanozyme	Pollutant	Enzyme-Like Activity	Year	Ref.
Fe_3O_4 MNP/MWNT composites	Phenol	Peroxidase	2009	[186]
$Fe_2(MoO_4)_3$	Acid Orange II	Peroxidase	2011	[190]
G-FeOOH/(reduced graphene oxide)	Phenol	Peroxidase	2011	[187]
Fe_3O_4/CeO_2 nanocomposites	4-Chlorophenol	Peroxidase	2012	[158]
Fe_3O_4/MSU-F-C (magnetite-loaded mesocellular carbonaceous material)	Arsenic, phenol	Peroxidase	2012	[188]
Fe_xO_y-MWNT	Orange G	Peroxidase	2012	[199]
MIL-101 (chromium(III) terephthalate metal organic framework) (chelated to N,N-dimethylamino pyridine)	Paraoxon	Hydrolase	2013	[197]
Porous Co_3O_4 nanorods–reduced graphene oxide	Methylene blue	Peroxidase	2013	[191]
Fe_3O_4/reduced graphene oxide nanocomposites	Methylene blue	Peroxidase	2013	[200]
graphene oxide quantum dots/Fe_3O_4 composites	Phenolic compounds	Peroxidase	2014	[48]
graphene oxide/Fe_3O_4 nanocomposites	Acid Orange 7	Peroxidase	2014	[192]
Polyprrole/hemin (PPy/hemin) nanocomposite	Methyl orange and Rh B	Pollutant adsorbent (not enzymatic activity)	2014	[193]
Fe_3O_3/graphene oxide	Rh B	Peroxidase	2014	[201]
CeO_2/G-Fe_2O_3	Parathion-methyl, nerve agents soman and VX	Peroxidase	2015	[198]
Fe_3O_4/Al-B (Fe_3O_4 MNPs decorated Al pillared bentonite)	Rh B	Peroxidase + adsorption	2015	[194]
Fe_3O_4/CNTs	Orange II	Peroxidase	2016	[195]
C-CoM-HNCs (HNC = homobimetallic hollow cages), (C-CoM-HNC M = Ni, Mn, Cu and Zn)	Rh B	Oxidase	2019	[196]
DhH6-c-ZrMOF (deterohemin-peptide on Zr metal organic framework)	Phenol	Peroxidase	2020	[189]
Cellulose incorporated magnetic nano-biocomposites, Fe_3O_4 on cellulose	Methyl orange, textile effluent	Peroxidase	2020	[159]

Different kinds of carbon materials have been investigated as nanocomposites with metal/metal oxide nanoparticles. Examples include carbon nanotubes, multiwalled carbon nanotubes (MWNTs), graphene, graphene quantum dots, graphene oxide, mesoporous carbon, etc. [202]. Zuo et al. [186] synthesised Fe_3O_4 MNP/MWNT nanocomplexes with peroxidase-like activity. In the presence of H_2O_2, the nanocomplex could efficiently catalyse the oxidative degradation of phenols to insoluble polyaromatic products that could be easily separated from aqueous solutions. Magnetically recoverable Fe_xO_y-MWNT was active in the degradation of Orange G dye [199]. The Fe_3O_4 nanoparticles (7 nm in size) that were grown on carbon nanotubes were shown to have higher catalytic reactivity as an enzyme mimic for the decomposition of the dye Orange II than Fe_3O_4 nanoparticles [195].

Peng [187] reported the synthesis of graphene-templated formation of ultrathin (2.1 nm) 2D lepidocrocite G-FeOOH and showed that these nanostructures could catalyse the degradation of phenol in the presence of H_2O_2. Nanocomposites between graphene oxide quantum dots (graphene sheets with lateral sizes less than 100 nm) and Fe_3O_4 nanoparticles were also shown to be effective for the degradative removal of phenolic compounds via peroxidase-like reactivity [48]. The high reactivity of the nanocomposite (higher than HRP) was attributed to the unique properties of the graphene oxide quantum dots, namely high electron conjugation and better aqueous dispersion ability compared to graphene oxide sheets, and the synergistic interactions between graphene oxide quantum dots and Fe_3O_4 MNPs. Fe_3O_4/reduced graphene oxide [200] and Fe_3O_3/graphene oxide [201] were reported to be efficient for the degradation of methylene blue and Rhodamine B at neutral pH, respectively. In a report by Zubir and co-workers [192], graphene oxide–Fe_3O_4 nanocomposites were found to be efficient in the degradation of acid orange dye. In comparison with Fe_3O_4 MNPs, the Fe_3O_3/graphene oxide had similar reactivity in the first 45 min of the catalytic cycle. However, while Fe_3O_4 was completely deactivated after one use, the nanocomposites retained good reactivity for the entire course of the reaction.

Porous Co_3O_4 nanorod-reduced graphene oxide (PCNG) (Figure 22) was shown to have high peroxidase activity in the degradation of methylene blue [191]. The improved catalytic activity of the nanocomposites could be the result of the synergy between the functions of porous Co_3O_4 nanorods and reduced graphene oxides. Here, due to the π-π stacking between methylene blue and the aromatic areas of reduced graphene oxide, methylene blue was more easily adsorbed on the surface of the PCNG. Electrons from methylene blue are donated to the PCNG, which leads to an increase in electron density and mobility in the PCNG. Electron transfer from PCNG to H_2O_2 is accelerated, thus increasing the reaction rate of methylene blue oxidation by H_2O_2. The PCNG had high thermal stability and was stable in the presence of an organic solvent, e.g., ethanol, tetrahydrofuran and N,N-dimethylformamide.

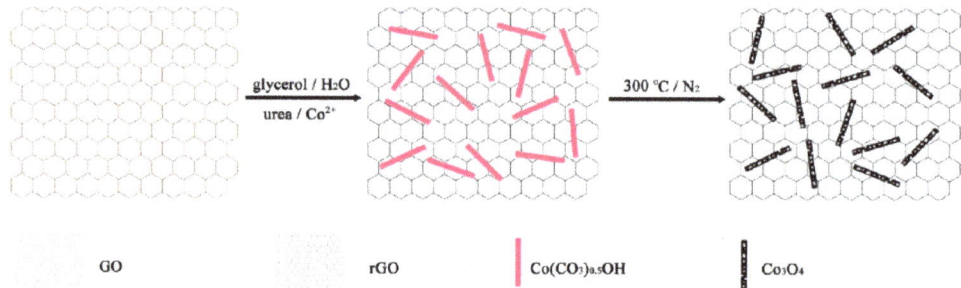

Figure 22. Procedure used in the preparation of porous Co_3O_4 nanorod-reduced graphene oxide, PCNG. (Reprinted with permission from [191]. Copyright © 2013, American Chemical Society).

Chun et al. [188] showed that magnetite (Fe_3O_4)-loaded meso-cellular carbonaceous material, Fe_3O_4/MSU-F-C, was very efficient in the Fenton-like reaction as well as an adsorbent for the removal of phenol and arsenic (Figure 23). The material could be easily separable by applying a magnetic field due to its strong magnetic property.

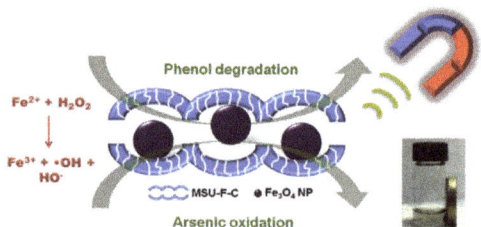

Figure 23. Removal of arsenic and phenol using Fe_3O_4/MSU-F-C as a catalytic adsorbent. (Reprinted with permission from [188]. Copyright © 2012 Elsevier Ltd. All rights reserved).

In short, the increase in the performance of nanocomposites between magnetically separable iron species (and other metal species) and carbon-based materials was attributed to several synergistic effects which include: (i) the pollutant molecules are brought closer to the active sites by the increased adsorptive interactions of the carbon phase; at the active sites, strongly oxidising HO• radicals are generated and react with the pollutant. The HO• radical would have lower probability of partaking in the non-productive parasitic reaction with H_2O_2, thus increasing the efficiency of H_2O_2 consumption; (ii) iron–carbon nanocomposites usually have good structural stability and lower leaching of metal species due to the confinement effect imposed by the carbon phase; (iii) the regeneration of the active sites is enhanced either by electron transfer features or delocalisation of π electrons of the carbon-based materials; (iv) the active sites are highly dispersed as the result of the high specific area of the carbon phase; and (v) some carbon materials also have peroxidase-like activity on their own [202]. Several of these nanocomposites have been shown to be effective for environmental pollutant degradation at neutral pH [199–202], which is a significant advantage as the treated water (in a wastewater treatment plant) would not be required to undergo a neutralisation step.

In addition to carbon materials, other materials have also been utilised in the preparation of nanocomposites for environmental pollutant degradation. Wan et al. [194] showed that Fe_3O_4 nanoparticle-decorated Al-pillared bentonite (Fe_3O_4/Al-B) had higher ability for the adsorption and degradation of Rh B than bare Fe_3O_4 (Figure 24). The composite also showed high stability and could be conveniently recycled.

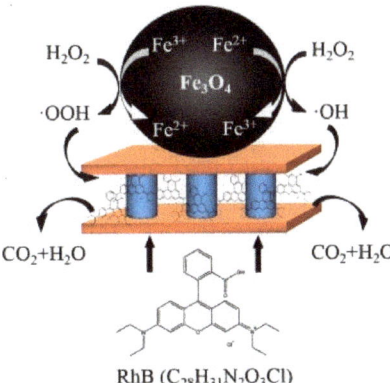

Figure 24. Fe_3O_4-decorated Al-pillared bentonite (Fe_3O_4/Al-B) as a peroxidase-like nanozyme for the degradation of Rh B. (Reprinted with permission from [194]. Copyright © 2015 Elsevier B.V. All rights reserved).

Fe_3O_4/CeO_2 nanocomposites (5–10 nm in size) were reported by Xu and co-workers [158] to be effective for the degradation of 4-chlorophenol at acidic pH of 3 in the presence of H_2O_2 via the peroxidase-like activity. 4-Chlorophenol was decomposed by the $HO^\bullet$ radical including surface-bound and in-solution radicals. The material could be reused several times. The proposed mechanism for the generation of hydroxyl radical $HO^\bullet$ from H_2O_2 and the interactions between different oxidation states of iron and cerium ions to facilitate $HO^\bullet$ formation is given in Figure 25. Janoš et al. [198] synthesised magnetically separable composites consisting of Fe_2O_3 grains and CeO_2 nanocrystalline surface (CeO_2/c-Fe_2O_3) and applied them as a "reactive sorbent" for the decomposition of dangerous organophosphorus compounds including organophosphorus pesticide parathion-methyl.

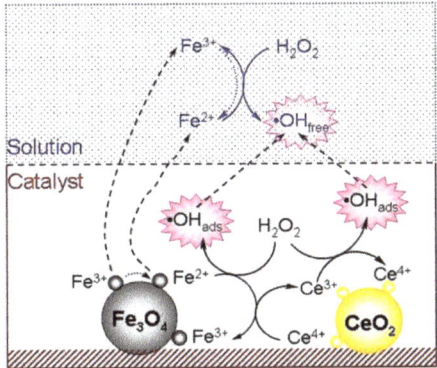

Figure 25. Proposed reaction mechanism of the H_2O_2 activation by Fe_3O_4/CeO_2 catalyst under acidic pH. (Reprinted with permission from [158]. Copyright © 2012, American Chemical Society).

Deuterohemin-peptide conjugated onto metal–organic framework, DhHP-c-ZrMOF, acted as a peroxidase mimetic catalyst for the degradation of phenol [189]. The schematic summary of the synthetic process is outlined in Figure 26. It was found that the hemin attached onto the MOF was significantly more active for the degradation of phenol and could operate in a wider pH range.

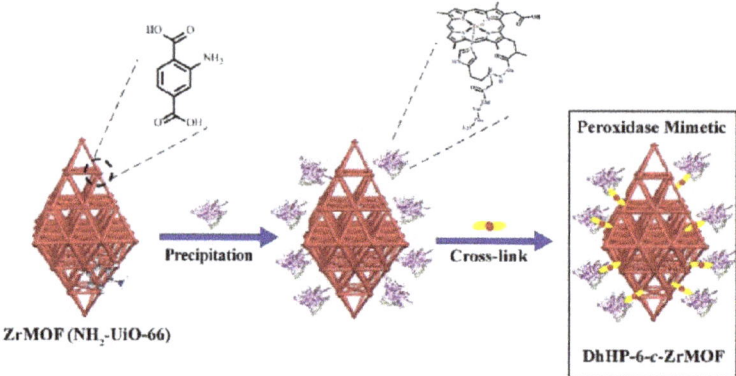

Figure 26. Schematic illustration of the synthesis of DhHP-c-ZrMOF. (Reprinted with permission from [189]. Copyright © 2019 Elsevier B.V. All rights reserved).

Chromium(III) terephthalate metal organic framework (MIL-101) was demonstrated to be effective in facilitating organophosphorus ester degradation using paraoxon as the model substrate [197]. The MIL-101 had an optimal activity at a basic pH of 10. The degradation of paraoxon (Figure 27) generates two acidic products which are likely to be removed in basic pH by the reaction with hydroxyl functional group.

Figure 27. MIL-101 with chelated aminopyridines as a catalyst for organophosphorus ester degradation. (Reprinted with permission from [197]. Copyright © 2013, American Chemical Society).

Li and co-workers [196] showed that the active sites in MOF-derived homobimetallic hollow nanocages are highly efficient as multifunctional nanozyme catalysts for biosensing and organic pollutant degradation. They fabricated Co-based homobimetallic hollow nanocages (HNCs)(C–CoM–HNC, M = Ni, Mn, Cu and Zn) and showed that the material acted as an efficient nanozyme for biosensing based on the excellent oxidase-like activity and for the degradation of an organic pollutant (Rh B).

Examples of other types of composite nanozymes were also reported. Polypyrrole/hemin nanocomposites were prepared and shown to have biosensing, dye removal ability and photothermal therapy [193]. The dye removal ability is not related to the catalytic activity of the nanocomposite and was established by the author to be an adsorbent effect. Cellulose-incorporated iron oxide magnetic nano-biocomposites as a peroxidase mimic have the potential to be a low-cost, recyclable option for the remediation of textile dyes, using the azo dye methyl orange as well as a textile effluent [159]. Niu et al. [203] prepared alginate/Fe@Fe$_3$O$_4$ core/shell structured nanoparticles (Fe$_3$O$_4$@ALG/Fe MNPs) for the defluorination and removal of norfloxacin (a fluoroquinolone antibiotic). The Fe$_3$O$_4$@ALG/Fe MNPs have higher efficiency for norfloxacin degradation compared with the Fe$_3$O$_4$ nanoparticle–H$_2$O$_2$ system, with 100% of the norfloxacin removed within 60 min. Mixed metal oxide, iron molybdate (Fe$_2$(MoO$_4$)$_3$), was prepared and shown to act as a heterogeneous Fenton-like catalyst for the degradation of the dye Acid Orange II. The heterogeneous catalyst worked efficiently in a relatively wide pH range of 3 to 9. Good mineralisation of Acid Orange II was achieved [190].

5.5. Type IV Nanozymes: 3D Structural Mimics

There are only a limited number of studies using type IV nanozymes for the degradation of environmental pollutants (Table 6). The technical challenge in the construction of these 3D nanostructures is likely to be a contributing factor to this limitation.

Table 6. Examples of type IV nanozymes in environmental pollutant remediation.

Nanozyme	Pollutant	Enzyme-Like Activity	Year	Ref.
Porous Fe$_3$O$_4$ nanospheres	Xylenol Orange	Peroxidase	2011	[204]
3D nano-assembly of Au nanoparticles on Au@Ag@ICPs	TMB and methylene blue	Oxidase	2015	[205]
Iron single-atom nanozyme, FeN$_4$	TMB and OPD	Peroxidase, oxidase and catalase	2019	[206]

Zhu and Dao [204] prepared porous Fe$_3$O$_4$ nanospheres (Figure 28) and showed that they were highly effective as catalysts for the degradation of xylenol orange with H$_2$O$_2$ as the oxidant in aqueous solution. The porous Fe$_3$O$_4$ could be recycled multiple (7) times, with only a slight drop in activity after each cycle.

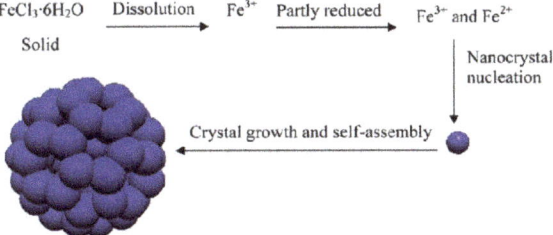

Figure 28. Preparation of porous Fe$_3$O$_4$ nanospheres. (Reprinted with permission from [204]. Copyright © 2011, American Chemical Society).

Wang et al. [205] fabricated three-dimensional nano-assemblies of noble metal nanoparticle (NP)–infinite coordination polymers (ICPs) through the infiltration of HAuCl$_4$ into hollow Au@Ag@ICPs core-shell nanostructures and its replacement reaction with Au@Ag nanoparticles (Figure 29). These 3D nano-assemblies possess specific oxidase-like activity. TMB was oxidised to generate its blue product using surface-adsorbed O$_2$ on the surface without using H$_2$O$_2$. Methylene blue was also degraded using the oxidase-like activity of the 3D assemblies.

Porphyrin-like single Fe sites on N-doped carbon nanomaterials (iron single-atom nanozymes, FeN$_4$) were constructed by Zhao and co-workers [206] using highly specialised high-temperature techniques (Figure 30). These iron single-atom nanozymes exhibited excellent peroxidase-, oxidase- and catalase-like activities. The catalytic activities could be up 40 times higher than those of Fe$_3$O$_4$. The enhanced reactivity could be attributed to the FeN$_4$ sites, which are similar to the natural heme-containing enzymes, the high surface area of the MOF (ZIF-8) structure and the large pore diameter (0.45 nm). These last two factors are beneficial for the mass transfer of reactant and product. TMB and OPD were the two substrates used for their colorimetric investigation, which showed promising results. The material was effective in the degradation of phenol in aqueous solution.

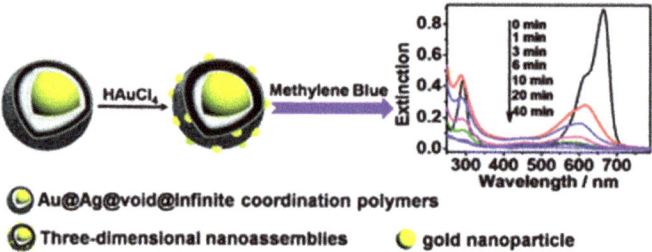

Figure 29. Three-dimensional assembly of gold nanoparticles on Au@Ag@void@infinite coordination polymers with oxidase-like activity for the degradation of methylene blue. (Reprinted with permission from [205]. Copyright © 2014, Royal Society of Chemistry).

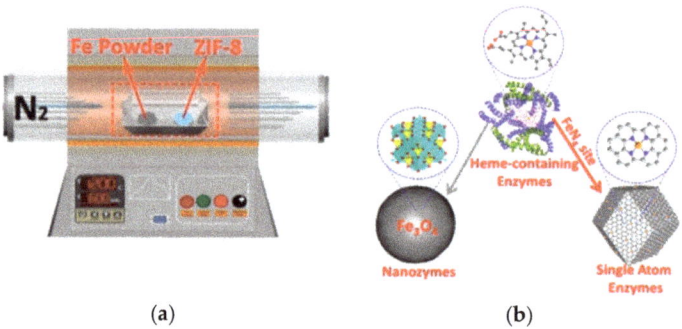

Figure 30. (a) Schematic illustration of the synthesis of iron single-atom nanozymes and (b) macrostructures and active sites of natural enzymes, nanozymes and iron single-atom nanozymes. (Reprinted with permission from [206]. Copyright © 2019, Royal Society of Chemistry).

5.6. Potential Applications of Nanozymes in the Treatment of PFAS (Per- and Polyfluoroalkyl Substances) as Emerging Pollutants

Per- and polyfluoroalkyl substances, PFAS, are a large group of chemicals which have been used extensively in many commercial products including fire-fighting foams, lubricants, coatings, etc. These chemicals can travel great distances in the environment and have been known to be bioaccumulative. Although the toxicity of these compounds has not been fully understood, they are suspected to be the cause of carcinogenesis, mutagenesis and reproductive problems. They are now considered as emerging pollutants [207,208]. A number of technologies have emerged for the remediation of PFAS pollutants; however, the high stability of C-F bonds presents a great challenge in devising an effective method for the degradation of PFAS under mild conditions [207–209]. Recent works have shown that microbial and enzymes could affect the degradation of certain PFAS compounds. HRP successfully catalysed the degradation of perfluorooctanoic acid [210] and laccases could also catalyse the degradation of perfluorooctanoic acid [211]. Given the fact that many nanozymes have been demonstrated to have peroxidase and laccase activities as well have significant advantages over natural enzymes, nanozymes could play an important role in PFAS pollutant remediation efforts.

5.7. Summary

The peroxidase-like catalytic activity has been the activity most explored for the degradation of environmental pollutants such as phenols, Rh B, dyes (methylene blue, acid orange), etc. Nanozymes have several advantages including low cost, ease of preparation, high stability and recyclability. Zero valent metal-based nanozymes have been least utilised as catalysts for environmental pollutant remediation. Noble metals are expensive for large-scale applications and zero valent non-noble metals tend to be highly prone to oxidation and degradation. Moreover, 3D-structured materials that possess high specificity (moulded in active sites) have great potential in disease diagnosis. However, in pollutant degradation, high selectivity and specificity is often not required. Additionally, the manufacturing of 3D structures still requires significant technical efforts.

Composite nanozymes, particularly those with Fe_3O_4 MNPs nanoparticles on carbon materials or MOFs, have been shown to exhibit higher catalytic efficiency than metal-/metal-oxide-based nanozymes. One of the attractive features of these nanocomposites is that they have dual adsorption and degradation activities. The adsorptive ability of the base material, i.e., carbon or MOF, could bring the pollutant to the proximity of the active sites and increase the degradation efficiency of the composite nanozyme.

Most nanozymes were only efficient in the degradation of environmental pollutants under acidic pH conditions. This means that in wastewater treatment application, the treated wastewater will need to undergo an additional and potentially costly neutralisation step. Finding nanozymes that can degrade pollutants under neutral conditions would remarkably boost the application of nanozymes in this area.

6. Conclusions and Future Perspectives

Nanozymes have gathered increasing research interest since their first discovery less than 15 years ago [6] because they exhibit catalytic properties and offer improved tolerance to harsh conditions compared to natural enzymes. They have demonstrated applications in the biomedical and environmental fields, including diagnostics and therapeutics, sensing, environmental monitoring and remediation of environmental pollutants. In this review, we detailed the classification of nanozymes, their general catalytic mechanisms, as well as their recent progress in environmental applications through discriminating their diverse detection and remediation platforms.

Public concerns about environmental safety call for innovative and informative analytical techniques to meet the (i) detection requirements of high sensitivity and specificity, and (ii) remediation requirements of degradation efficiency to transform pollutants into another form less toxic to the environment. Nanozymes have emerged as an excellent tool to address both environmental detection and remediation requirements. Recent works have shown that nanozymes have fascinating catalytic properties, with added advantages over natural enzymes for environmental applications. Not only do they have wider operational windows (resistance to harsh environments), higher stability (long shelf life) and better recyclability, but they also offer tuneable surface functionality. The highly dynamic and active research has given rise to great opportunities in this field. It is believed that in certain applications, the combination of nanozymes with natural enzymes may lead to positive synergistic effects. However, an in-depth understanding of the fundamental principles of nanozymes for environmental quality and safety detection and remediation remains limited, which makes their applications largely empirical. To support sustainable growth and to realise the implementation of nanozymes in the environmental field, there is still scope for improvement.

1. Selectivity: Natural enzymes often have a defined size and morphology and catalyse a specific substrate or a class of analogues. However, nanozymes do not possess as high a substrate (target) selectivity as natural enzymes and their catalytic behaviours can be influenced by ions in the microenvironment, particularly anions. Consequently, more work in innovative surface engineering of nanozymes is still required to create target-tuneable catalysis which can perform selective recognition of the target.

2. Real-time in-field application: To meet the goal of real-time in-field detection of environmental pollutants, further studies are required to minimise sample pre-treatment (such as sample filtration, pH adjustment, solvent extraction, etc.). The vast majority of nanozymes for environmental pollutant monitoring and remediation are based on peroxidase-like activity, which has two drawbacks: (i) they require the addition of unstable hydrogen peroxide, and (ii) acidic pH adjustment to enhance signal. Ideally, nanozymes that can operate under neutral pH conditions without cumbersome sample modification can make the test more user-friendly. Furthermore, implementation of new and/or existing nanozymes into portable devices such as lateral flow assays and microfluidic chambers, coupled with the ability to capture data using portable optical readers such as smartphone cameras (for colorimetric detection), and combined with the fast data processing capability of a smartphone, will help to realise this goal.
3. Effectiveness and reusability of nanozymes: Nanocomposites between MNPs and carbon materials or MOFs have been shown to be remarkably more effective in the degradation of organic pollutants. The improved efficiencies are a result of the combined adsorption and catalytic abilities of the base materials and the MNPs and synergistic interactions between the two materials. Furthermore, while most nanozymes have optimal catalytic activity at acidic pH, examples of composite nanozymes which are highly active at neutral pH are also known. Accordingly, further studies into the development of nanozymes that are highly active at neutral pH will be beneficial for both the detection and the remediation of pollutants as this will reduce sample pre- and post-treatments, e.g., no acidification or post-neutralisation steps are required. Additionally, nanozymes that can be regenerated are also helpful, particularly in the remediation of environmental pollutants (such that they can continue to degrade more pollutants). Thus, reusability and ecological compatibility are also valuable considerations when designing new remediation strategies.
4. Commercial production: Currently, most nanoparticles used in research studies are synthesised in small batches using methods which are usually labour-intensive and time-consuming. Since nanoparticles have a large surface area to volume (and mass) ratio with greater reactivity and mobility, they have the tendency to agglomerate into larger microparticles, losing their distinctive nano characteristics. Small synthetic batches are also prone to batch-to-batch variability and a wide particle size distribution, which can significantly affect the yield of production. While small batches are adequate for early studies, this limits the translation of promising nanozymes for commercial deployment in environmental applications, where orders of magnitude more material are required. Development of methodologies such as flow chemistry techniques that can produce gram quantities per hour of highly reproducible nanozymes, combined with coating strategies to reduce agglomeration at the commercial scale, is important to support scale-up production.
5. Industrial standards and regulation: There is no doubt that nanozymes provide industries with many advantages. However, there are concerns related to the impact of nanoparticles on the ecological system, especially when deployed at large scale for remediation. Current regulations establish metal content limits without consideration of particle size. While implementing nanozymes as an environmental remediation tool, the development should be coupled with appropriate measurement techniques that can quantify both concentrations and particle sizes with appropriate quality assurance and quality control. As well as size and composition, it is evident that the surface properties of nanoparticles will be fundamental in determining the fate and toxicity in the environment and that these properties will need to be considered in any hazard ranking. Consequently, it is necessary to develop computational models to correlate the physicochemical properties of such nanozymes with their potential nanotoxicity. These models can also support public needs and industrial regulations for future remediation designs.

Author Contributions: Writing—original draft preparation, E.L.S.W., K.Q.V., E.C.; writing—review and editing, E.L.S., K.Q.V., E.C. All authors have read and agreed to the published version of the manuscript.

Funding: This research received no external funding.

Institutional Review Board Statement: Not applicable.

Informed Consent Statement: Not applicable.

Data Availability Statement: Data sharing not applicable.

Conflicts of Interest: The authors declare no conflict of interest.

Abbreviations

AAS	atomic absorption spectroscopy
ABTS	2,2′-azino-bis-(3-ethylbenzothiazoline-6-sulfonate)
AChE	acetylcholinesterase
BPA	bisphenol A
CHO	choline oxidase
EDTA	ethylenediaminetetraacetic acid
GC-MS	gas chromatography–mass spectrometry
HRP	horseradish peroxidase
ICP-MS	inductively coupled plasma–mass spectrometry
ICP-OES	inductively coupled plasma–optical emission spectroscopy
LC-MS	liquid chromatography–mass spectrometry
MNP	magnetic nanoparticles
MOF	metal–organic framework
MWNT	multiwalled carbon nanotube
OPD	o-phenylenediamine
PFAS	per- and polyfluoroalkyl substances
Rh B	Rhodamine B
TMB	3,3′,5,5′-tetramethylbenzidine

References

1. Jiang, D.W.; Ni, D.L.; Rosenkrans, Z.T.; Huang, P.; Yan, X.Y.; Cai, W.B. Nanozyme: New horizons for responsive biomedical applications. *Chem. Soc. Rev.* **2019**, *48*, 3683–3704. [CrossRef]
2. Wang, Q.Q.; Wei, H.; Zhang, Z.Q.; Wang, E.K.; Dong, S.J. Nanozyme: An emerging alternative to natural enzyme for biosensing and immunoassay. *Trends Anal. Chem.* **2018**, *105*, 218–224. [CrossRef]
3. Huang, Y.; Ren, J.; Qu, X. Nanozymes: Classification, catalytic mechanisms, activity regulation, and applications. *Chem. Rev.* **2019**, *119*, 4357–4412. [CrossRef]
4. Wang, X.Y.; Hu, Y.H.; Wei, H. Nanozymes in bionanotechnology: From sensing to therapeutics and beyond. *Inorg. Chem. Front.* **2016**, *3*, 41–60. [CrossRef]
5. Gao, Y.; Zhou, Y.Z.; Chandrawati, R. Metal and metal oxide nanoparticles to enhance the performance of enzyme-linked immunosorbent assay (ELISA). *ACS Appl. Nano Mater.* **2020**, *3*, 1–21. [CrossRef]
6. Gao, L.; Zhuang, J.; Nie, L.; Zhang, J.; Zhang, Y.; Gu, N.; Wang, T.; Feng, J.; Yang, D.; Perrett, S. Intrinsic peroxidase-like activity of ferromagnetic nanoparticles. *Nat. Nanotechnol.* **2007**, *2*, 577–583. [CrossRef]
7. Gao, Z.Q.; Xu, M.D.; Hou, L.; Chen, G.N.; Tang, D.P. Magnetic bead-based reverse colorimetric immunoassay strategy for sensing biomolecules. *Anal. Chem.* **2013**, *85*, 6945–6952. [CrossRef]
8. Wang, Z.F.; Zheng, S.; Cai, J.; Wang, P.; Feng, J.; Yang, X.; Zhang, L.M.; Ji, M.; Wu, F.G.; He, N.Y.; et al. Fluorescent artificial enzyme-linked immunoassay system based on Pd/C nanocatalyst and fluorescent chemodosimeter. *Anal. Chem.* **2013**, *85*, 11602–11609. [CrossRef]
9. Dong, J.L.; Song, L.N.; Yin, J.J.; He, W.W.; Wu, Y.H.; Gu, N.; Zhang, Y. Co_3O_4 nanoparticles with multi-enzyme activities and their application in immunohistochemical assay. *ACS Appl. Mater. Interfaces* **2014**, *6*, 1959–1970. [CrossRef]
10. Gao, Z.Q.; Xu, M.D.; Lu, M.H.; Chen, G.N.; Tang, D.P. Urchin-like (gold core)@(platinum shell) nanohybrids: A highly efficient peroxidase-mimetic system for in situ amplified colorimetric immunoassay. *Biosens. Bioelectron.* **2015**, *70*, 194–201. [CrossRef]
11. Huo, M.F.; Wang, L.Y.; Chen, Y.; Shi, J.L. Tumor-selective catalytic nanomedicine by nanocatalyst delivery. *Nat. Commun.* **2017**, *8*, 357. [CrossRef] [PubMed]

12. Wang, G.L.; Xu, X.F.; Qiu, L.; Dong, Y.M.; Li, Z.J.; Zhang, C. Dual responsive enzyme mimicking activity of AgX (X = Cl, Br, I) nanoparticles and its application for cancer cell detection. *ACS Appl. Mater. Interfaces* **2014**, *6*, 6434–6442. [CrossRef] [PubMed]
13. Maji, S.K.; Mandal, A.K.; Nguyen, K.T.; Borah, P.; Zhao, Y.L. Cancer cell detection and therapeutics using peroxidase-active nanohybrid of gold nanoparticle-loaded mesoporous silica-coated graphene. *ACS Appl. Mater. Interfaces* **2015**, *7*, 9807–9816. [CrossRef] [PubMed]
14. Tian, Z.M.; Li, J.; Zhang, Z.Y.; Gao, W.; Zhou, X.M.; Qu, Y.Q. Highly sensitive and robust peroxidase-like activity of porous nanorods of ceria and their application for breast cancer detection. *Biomaterials* **2015**, *59*, 116–124. [CrossRef] [PubMed]
15. Asati, A.; Kaittanis, C.; Santra, S.; Perez, J.M. pH-tunable oxidase-like activity of cerium oxide nanoparticles achieving sensitive fluorigenic detection of cancer biomarkers at neutral pH. *Anal. Chem.* **2011**, *83*, 2547–2553. [CrossRef]
16. Wei, H.; Wang, E.K. Nanomaterials with enzyme-like characteristics (nanozymes): Next-generation artificial enzymes. *Chem. Soc. Rev.* **2013**, *42*, 6060–6093. [CrossRef]
17. Wu, J.J.X.; Wang, X.Y.; Wang, Q.; Lou, Z.P.; Li, S.R.; Zhu, Y.Y.; Qin, L.; Wei, H. Nanomaterials with enzyme-like characteristics (nanozymes): Next-generation artificial enzymes (II). *Chem. Soc. Rev.* **2019**, *48*, 1004–1076. [CrossRef]
18. Kazlauskas, R.J. Enhancing catalytic promiscuity for biocatalysis. *Curr. Opin. Chem. Biol.* **2005**, *9*, 195–201. [CrossRef]
19. Khersonsky, O.; Roodveldt, C.; Tawfik, D.S. Enzyme promiscuity: Evolutionary and mechanistic aspects. *Curr. Opin. Chem. Biol.* **2006**, *10*, 498–508. [CrossRef]
20. Holm, R.H.; Kennepohl, P.; Solomon, E.I. Structural and functional aspects of metal sites in biology. *Chem. Rev.* **1996**, *96*, 2239–2314. [CrossRef]
21. Palizban, A.A.; Sadeghi-Aliabadi, H.; Abdollahpour, F. Effect of cerium lanthanide on Hela and MCF-7 cancer cell growth in the presence of transferrin. *Res. Pharm. Sci.* **2010**, *5*, 119–125. [PubMed]
22. Celardo, I.; Pedersen, J.Z.; Traversa, E.; Ghibelli, L. Pharmacological potential of cerium oxide nanoparticles. *Nanoscale* **2011**, *3*, 1411–1420. [CrossRef] [PubMed]
23. Baldim, V.; Bedioui, F.; Mignet, N.; Margaill, I.; Berret, J.F. The enzyme-like catalytic activity of cerium oxide nanoparticles and its dependency on Ce^{3+} surface area concentration. *Nanoscale* **2018**, *10*, 6971–6980. [CrossRef] [PubMed]
24. Mu, J.S.; Wang, Y.; Zhao, M.; Zhang, L. Intrinsic peroxidase-like activity and catalase-like activity of Co_3O_4 nanoparticles. *Chem. Commun.* **2012**, *48*, 2540–2542. [CrossRef] [PubMed]
25. Lin, S.B.; Wang, Y.Y.; Chen, Z.Z.; Li, L.B.; Zeng, J.F.; Dong, Q.R.; Wang, Y.; Chai, Z.F. Biomineralized enzyme-like cobalt sulfide nanodots for synergetic phototherapy with tumor multimodal imaging navigation. *ACS Sustain. Chem. Eng.* **2018**, *6*, 12061–12069. [CrossRef]
26. Chen, W.; Chen, J.; Liu, A.L.; Wang, L.M.; Li, G.W.; Lin, X.H. Peroxidase-like activity of cupric oxide nanoparticle. *ChemCatChem.* **2011**, *3*, 1151–1154. [CrossRef]
27. Wan, Y.; Qi, P.; Zhang, D.; Wu, J.J.; Wang, Y. Manganese oxide nanowire-mediated enzyme-linked immunosorbent assay. *Biosens. Bioelectron.* **2012**, *33*, 69–74. [CrossRef]
28. Pijpers, I.A.B.; Cao, S.; Llopis-Lorente, A.; Zhu, J.; Song, S.; Joosten, R.R.M.; Meng, F.; Friedrich, H.; Williams, D.S.; Sánchez, S.; et al. Hybrid biodegradable nanomotors through compartmentalized synthesis. *Nano Lett.* **2020**, *20*, 4472–4480. [CrossRef]
29. He, W.W.; Zhou, Y.T.; Warner, W.G.; Hu, X.N.; Wu, X.C.; Zheng, Z.; Boudreau, M.D.; Yin, J.J. Intrinsic catalytic activity of Au nanoparticles with respect to hydrogen peroxide decomposition and superoxide scavenging. *Biomaterials* **2013**, *34*, 765–773. [CrossRef]
30. Liu, C.P.; Wu, T.H.; Lin, Y.L.; Liu, C.Y.; Wang, S.; Lin, S.Y. Tailoring enzyme-like activities of gold nanoclusters by polymeric tertiary amines for protecting neurons against oxidative stress. *Small* **2016**, *12*, 4127–4135. [CrossRef]
31. Long, R.; Huang, H.; Li, Y.P.; Song, L.; Xiong, Y.J. Palladium-based nanomaterials: A platform to produce reactive oxygen species for catalyzing oxidation reactions. *Adv. Mater.* **2015**, *27*, 7025–7042. [CrossRef] [PubMed]
32. Comotti, M.; Della Pina, C.; Matarrese, R.; Rossi, M. The catalytic activity of "naked" gold particles. *Angew. Chem. Int. Ed.* **2004**, *43*, 5812–5815. [CrossRef] [PubMed]

33. Wang, S.; Chen, W.; Liu, A.L.; Hong, L.; Deng, H.H.; Lin, X.H. Comparison of the peroxidase-like activity of unmodified, amino-modified, and citrate-capped gold nanoparticles. *ChemPhysChem* **2012**, *13*, 1199–1204. [CrossRef] [PubMed]
34. Jv, Y.; Li, B.X.; Cao, R. Positively-charged gold nanoparticles as peroxidiase mimic and their application in hydrogen peroxide and glucose detection. *Chem. Commun.* **2010**, *46*, 8017–8019. [CrossRef] [PubMed]
35. Li, J.N.; Liu, W.Q.; Wu, X.C.; Gao, X.F. Mechanism of pH-switchable peroxidase and catalase-like activities of gold, silver, platinum and palladium. *Biomaterials* **2015**, *48*, 37–44. [CrossRef] [PubMed]
36. Fan, J.; Yin, J.J.; Ning, B.; Wu, X.C.; Hu, Y.; Ferrari, M.; Anderson, G.J.; Wei, J.Y.; Zhao, Y.L.; Nie, G.J. Direct evidence for catalase and peroxidase activities of ferritin-platinum nanoparticles. *Biomaterials* **2011**, *32*, 1611–1618. [CrossRef]
37. Shen, X.M.; Liu, W.Q.; Gao, X.J.; Lu, Z.H.; Wu, X.C.; Gao, X.F. Mechanisms of oxidase and superoxide dismutation-like activities of gold, silver, platinum, and palladium, and their alloys: A general way to the activation of molecular oxygen. *J. Am. Chem. Soc.* **2015**, *137*, 15882–15891. [CrossRef]
38. Xu, Y.; Chen, L.; Wang, X.C.; Yao, W.T.; Zhang, Q. Recent advances in noble metal based composite nanocatalysts: Colloidal synthesis, properties, and catalytic applications. *Nanoscale* **2015**, *7*, 10559–10583. [CrossRef]
39. He, W.W.; Wu, X.C.; Liu, J.B.; Hu, X.N.; Zhang, K.; Hou, S.A.; Zhou, W.Y.; Xie, S.S. Design of AgM bimetallic alloy nanostructures (M = Au, Pd, Pt) with tunable morphology and peroxidase-like activity. *Chem. Mater.* **2010**, *22*, 2988–2994. [CrossRef]
40. He, W.W.; Liu, Y.; Yuan, J.S.; Yin, J.J.; Wu, X.C.; Hu, X.N.; Zhang, K.; Liu, J.B.; Chen, C.Y.; Ji, Y.L.; et al. Au@Pt nanostructures as oxidase and peroxidase mimetics for use in immunoassays. *Biomaterials* **2011**, *32*, 1139–1147. [CrossRef]
41. Xia, X.H.; Zhang, J.T.; Lu, N.; Kim, M.J.; Ghale, K.; Xu, Y.; McKenzie, E.; Liu, J.B.; Yet, H.H. Pd-Ir core-shell nanocubes: A type of highly efficient and versatile peroxidase mimic. *ACS Nano* **2015**, *9*, 9994–10004. [CrossRef] [PubMed]
42. Garg, B.; Bisht, T. Carbon nanodots as peroxidase nanozymes for biosensing. *Molecules* **2016**, *21*, 1653. [CrossRef] [PubMed]
43. Sun, H.; Zhou, Y.; Ren, J.; Qu, X. Carbon nanozymes: Enzymatic properties, catalytic mechanism, and applications. *Angew. Chem. Int. Ed.* **2018**, *57*, 9224–9237. [CrossRef] [PubMed]
44. Shi, W.; Wang, Q.; Long, Y.; Cheng, Z.; Chen, S.; Zheng, H.; Huang, Y. Carbon nanodots as peroxidase mimetics and their applications to glucose detection. *Chem. Commun.* **2011**, *47*, 6695–6697. [CrossRef]
45. Song, Y.J.; Qu, K.G.; Zhao, C.; Ren, J.S.; Qu, X.G. Graphene oxide: Intrinsic peroxidase catalytic activity and its application to glucose detection. *Adv. Mater.* **2010**, *22*, 2206–2210. [CrossRef]
46. Kim, M.S.; Cho, S.; Joo, S.H.; Lee, J.; Kwak, S.K.; Kim, M.I. N- and B-codoped graphene: A strong candidate to replace natural peroxidase in sensitive and selective bioassays. *ACS Nano* **2019**, *13*, 4312–4321. [CrossRef]
47. Ren, C.X.; Hu, X.G.; Zhou, Q.X. Graphene oxide quantum dots reduce oxidative stress and inhibit neurotoxicity in vitro and in vivo through catalase-like activity and metabolic regulation. *Adv. Sci.* **2018**, *5*, 1700595. [CrossRef]
48. Wu, X.C.; Zhang, Y.; Han, T.; Wu, H.X.; Guo, S.W.; Zhang, J.Y. Composite of graphene quantum dots and Fe$_3$O$_4$ nanoparticles: Peroxidase activity and application in phenolic compound removal. *RSC Adv.* **2014**, *4*, 3299–3305. [CrossRef]
49. Dong, Y.M.; Zhang, J.J.; Jiang, P.P.; Wang, G.L.; Wu, X.M.; Zhao, H.; Zhang, C. Superior peroxidase mimetic activity of carbon dots-Pt nanocomposites relies on synergistic effects. *New J. Chem.* **2015**, *39*, 4141–4146. [CrossRef]
50. Zheng, C.; Ke, W.J.; Yin, T.X.; An, X.Q. Intrinsic peroxidase-like activity and the catalytic mechanism of gold@carbon dots nanocomposites. *RSC Adv.* **2016**, *6*, 35280–35286. [CrossRef]
51. Chen, Q.M.; Liang, C.H.; Zhang, X.D.; Huang, Y.M. High oxidase-mimic activity of Fe nanoparticles embedded in an N-rich porous carbon and their application for sensing of dopamine. *Talanta* **2018**, *182*, 476–483. [CrossRef] [PubMed]
52. Farha, O.K.; Shultz, A.M.; Sarjeant, A.A.; Nguyen, S.T.; Hupp, J.T. Active-site-accessible, porphyrinic metal-organic framework materials. *J. Am. Chem. Soc.* **2011**, *133*, 5652–5655. [CrossRef] [PubMed]

53. Liu, Y.L.; Zhao, X.J.; Yang, X.X.; Li, Y.F. A nanosized metal-organic framework of Fe-MIL-88NH$_2$ as a novel peroxidase mimic used for colorimetric detection of glucose. *Analyst* **2013**, *138*, 4526–4531. [CrossRef] [PubMed]
54. Wang, C.H.; Gao, J.; Cao, Y.L.; Tan, H.L. Colorimetric logic gate for alkaline phosphatase based on copper (II)-based metal-organic frameworks with peroxidase-like activity. *Anal. Chim. Acta* **2018**, *1004*, 74–81. [CrossRef] [PubMed]
55. Chen, J.Y.; Shu, Y.; Li, H.L.; Xu, Q.; Hu, X.Y. Nickel metal-organic framework 2D nanosheets with enhanced peroxidase nanozyme activity for colorimetric detection of H$_2$O$_2$. *Talanta* **2018**, *189*, 254–261. [CrossRef] [PubMed]
56. Li, H.P.; Liu, H.F.; Zhang, J.D.; Cheng, Y.X.; Zhang, C.L.; Fei, X.Y.; Xian, Y.Z. Platinum nanoparticle encapsulated metal-organic frameworks for colorimetric measurement and facile removal of mercury(II). *ACS Appl. Mater. Interfaces* **2017**, *9*, 40716–40725. [CrossRef] [PubMed]
57. Yang, H.G.; Yang, R.T.; Zhang, P.; Qin, Y.M.; Chen, T.; Ye, F.G. A bimetallic (Co/2Fe) metal-organic framework with oxidase and peroxidase mimicking activity for colorimetric detection of hydrogen peroxide. *Microchim. Acta* **2017**, *184*, 4629–4635. [CrossRef]
58. Xu, H.M.; Liu, M.; Huang, X.D.; Min, Q.H.; Zhu, J.J. Multiplexed quantitative MALDI MS approach for assessing activity and inhibition of protein kinases based on postenrichment dephosphorylation of phosphopeptides by metal-organic framework-templated porous CeO$_2$. *Anal. Chem.* **2018**, *90*, 9859–9867. [CrossRef]
59. Zhang, W.; Hu, S.L.; Yin, J.J.; He, W.W.; Lu, W.; Ma, M.; Gu, N.; Zhang, Y. Prussian blue nanoparticles as multienzyme mimetics and reactive oxygen species scavengers. *J. Am. Chem. Soc.* **2016**, *138*, 5860–5865. [CrossRef]
60. Lin, Y.H.; Ren, J.S.; Qu, X.G. Catalytically active nanomaterials: A promising candidate for artificial enzymes. *Acc. Chem. Res.* **2014**, *47*, 1097–1105. [CrossRef]
61. Ghosh, S.; Roy, P.; Karmodak, N.; Jemmis, E.D.; Mugesh, G. Nanoisozymes: Crystal-facet-dependent enzyme-mimetic activity of V$_2$O$_5$ nanomaterials. *Angew. Chem. Int. Ed.* **2018**, *57*, 4510–4515. [CrossRef]
62. Huang, L.; Chen, J.X.; Gan, L.F.; Wang, J.; Dong, S.J. Single-atom nanozymes. *Sci. Adv.* **2019**, *5*, eaav5490. [CrossRef]
63. Benedetti, T.M.; Andronescu, C.; Cheong, S.; Wilde, P.; Wordsworth, J.; Kientz, M.; Tilley, R.D.; Schuhmann, W.; Gooding, J.J. Electrocatalytic nanoparticles that mimic the three-dimensional geometric architecture of enzymes: Nanozymes. *J. Am. Chem. Soc.* **2018**, *140*, 13449–13455. [CrossRef] [PubMed]
64. Chen, Y.J.; Ji, S.F.; Wang, Y.G.; Dong, J.C.; Chen, W.X.; Li, Z.; Shen, R.A.; Zheng, L.R.; Zhuang, Z.B.; Wang, D.S.; et al. Isolated single iron atoms anchored on N-doped porous carbon as an efficient electrocatalyst for the oxygen reduction reaction. *Angew. Chem. Int. Ed.* **2017**, *56*, 6937–6941. [CrossRef] [PubMed]
65. Huang, X.Y.; Groves, J.T. Oxygen activation and radical transformations in heme proteins and metalloporphyrins. *Chem. Rev.* **2018**, *118*, 2491–2553. [CrossRef] [PubMed]
66. Jiao, L.; Wu, J.B.; Zhong, H.; Zhang, Y.; Xu, W.Q.; Wu, Y.; Chen, Y.F.; Yan, H.Y.; Zhang, Q.H.; Gu, W.L.; et al. Densely isolated FeN$_4$ sites for peroxidase mimicking. *ACS Catal.* **2020**, *10*, 6422–6429. [CrossRef]
67. Wordsworth, J.; Benedetti, T.M.; Alinezhad, A.; Tilley, R.D.; Edwards, M.A.; Schuhmann, W.; Gooding, J.J. The importance of nanoscale confinement to electrocatalytic performance. *Chem. Sci.* **2020**, *11*, 1233–1240. [CrossRef]
68. Voinov, M.A.; Pagan, J.O.S.; Morrison, E.; Smirnova, T.I.; Smirnov, A.I. Surface-mediated production of hydroxyl radicals as a mechanism of iron oxide nanoparticle biotoxicity. *J. Am. Chem. Soc.* **2011**, *133*, 35–41. [CrossRef]
69. Fenton, H. LXXIII.—Oxidation of tartaric acid in presence of iron. *J. Chem. Soc. Trans.* **1894**, *65*, 899–910. [CrossRef]
70. Hu, X.N.; Saran, A.; Hou, S.; Wen, T.; Ji, Y.L.; Liu, W.Q.; Zhang, H.; He, W.W.; Yin, J.J.; Wu, X.C. Au@PtAg core/shell nanorods: Tailoring enzyme-like activities via alloying. *RSC Adv.* **2013**, *3*, 6095–6105. [CrossRef]
71. Zhang, H.J.; Okuni, J.; Toshima, N. One-pot synthesis of Ag-Au bimetallic nanoparticles with Au shell and their high catalytic activity for aerobic glucose oxidation. *J. Colloid. Interface Sci.* **2011**, *354*, 131–138. [CrossRef] [PubMed]
72. Nath, I.; Chakraborty, J.; Verpoort, F. Metal organic frameworks mimicking natural enzymes: A structural and functional analogy. *Chem. Soc. Rev.* **2016**, *45*, 4127–4170. [CrossRef]

73. Sun, H.J.; Zhao, A.D.; Gao, N.; Li, K.; Ren, J.S.; Qu, X.G. Deciphering a nanocarbon-based artificial peroxidase: Chemical identification of the catalytically active and substrate-binding sites on graphene quantum dots. *Angew. Chem. Int. Ed.* **2015**, *54*, 7176–7180. [CrossRef]
74. Zhao, R.S.; Zhao, X.; Gao, X.F. Molecular-level insights into intrinsic peroxidase-like activity of nanocarbon oxides. *Chem. Eur. J.* **2015**, *21*, 960–964. [CrossRef]
75. Cheng, N.; Li, J.C.; Liu, D.; Lin, Y.H.; Du, D. Single-atom nanozyme based on nanoengineered Fe-N-C catalyst with superior peroxidase-like activity for ultrasensitive bioassays. *Small* **2019**, *15*, 1901485. [CrossRef] [PubMed]
76. Xu, B.L.; Wang, H.; Wang, W.W.; Gao, L.Z.; Li, S.S.; Pan, X.T.; Wang, H.Y.; Yang, H.L.; Meng, X.Q.; Wu, Q.W.; et al. A single-atom nanozyme for wound disinfection applications. *Angew. Chem. Int. Ed.* **2019**, *58*, 4911–4916. [CrossRef] [PubMed]
77. Calle-Vallejo, F.; Tymoczko, J.; Colic, V.; Vu, Q.H.; Pohl, M.D.; Morgenstern, K.; Loffreda, D.; Sautet, P.; Schuhmann, W.; Bandarenka, A.S. Finding optimal surface sites on heterogeneous catalysts by counting nearest neighbors. *Science* **2015**, *350*, 185–189. [CrossRef]
78. *Australian Drinking Water Guidelines Paper 6 National Water Quality Management Strategy*; National Health and Medical Research Council, National Resource Management Ministerial Council: Canberra, Australia, 2011.
79. Olesik, J.W. Elemental analysis using ICP-OES and ICP/MS. *Anal. Chem.* **1991**, *63*, 12A–21A. [CrossRef]
80. Kaur, B.; Kaur, N.; Kumar, S. Colorimetric metal ion sensors—A comprehensive review of the years 2011–2016. *Coord. Chem. Rev.* **2018**, *358*, 13–69. [CrossRef]
81. Liu, B.; Zhuang, J.; Wei, G. Recent advances in the design of colorimetric sensors for environmental monitoring. *Environ. Sci.: Nano* **2020**, *7*, 2195–2213. [CrossRef]
82. Chow, E.; Gooding, J.J. Peptide modified electrodes as electrochemical metal ion sensors. *Electroanalysis* **2006**, *18*, 1437–1448. [CrossRef]
83. Bansod, B.; Kumar, T.; Thakur, R.; Rana, S.; Singh, I. A review on various electrochemical techniques for heavy metal ions detection with different sensing platforms. *Biosens. Bioelectron.* **2017**, *94*, 443–455. [CrossRef] [PubMed]
84. Priyadarshini, E.; Pradhan, N. Gold nanoparticles as efficient sensors in colorimetric detection of toxic metal ions: A review. *Sens. Actuators B Chem.* **2017**, *238*, 888–902. [CrossRef]
85. Johnson, R.D.; Bachas, L.G. Ionophore-based ion-selective potentiometric and optical sensors. *Anal. Bioanal. Chem.* **2003**, *376*, 328–341. [CrossRef] [PubMed]
86. Slocik, J.M.; Zabinski Jr, J.S.; Phillips, D.M.; Naik, R.R. Colorimetric response of peptide-functionalized gold nanoparticles to metal ions. *Small* **2008**, *4*, 548–551. [CrossRef]
87. Zhou, W.; Saran, R.; Liu, J. Metal sensing by DNA. *Chem. Rev.* **2017**, *117*, 8272–8325. [CrossRef]
88. Kukla, A.L.; Kanjuk, N.I.; Starodub, N.F.; Shirshov, Y.M. Multienzyme electrochemical sensor array for determination of heavy metal ions. *Sens. Actuators B Chem.* **1999**, *57*, 213–218. [CrossRef]
89. Krawczyński vel Krawczyk, T.; Moszczyńska, M.; Trojanowicz, M. Inhibitive determination of mercury and other metal ions by potentiometric urea biosensor. *Biosens. Bioelectron.* **2000**, *15*, 681–691. [CrossRef]
90. Unnikrishnan, B.; Lien, C.-W.; Chu, H.-W.; Huang, C.-C. A review on metal nanozyme-based sensing of heavy metal ions: Challenges and future perspectives. *J. Hazard. Mater.* **2021**, *401*, 123397. [CrossRef]
91. Li, X.; Wang, L.; Du, D.; Ni, L.; Pan, J.; Niu, X. Emerging applications of nanozymes in environmental analysis: Opportunities and trends. *Trends Anal. Chem.* **2019**, *120*, 115653. [CrossRef]
92. Meng, Y.; Li, W.; Pan, X.; Gadd, G.M. Applications of nanozymes in the environment. *Environ. Sci. Nano* **2020**, *7*, 135–1318. [CrossRef]
93. Yan, Z.; Yuan, H.; Zhao, Q.; Xing, L.; Zheng, X.; Wang, W.; Zhao, Y.; Yu, Y.; Hu, L.; Yao, W. Recent developments of nanoenzyme-based colorimetric sensors for heavy metal detection and the interaction mechanism. *Analyst* **2020**, *145*, 3173–3187. [CrossRef] [PubMed]
94. Zhao, Y.; Qiang, H.; Chen, Z.B. Colorimetric determination of Hg(II) based on a visually detectable signal amplification induced by a Cu@Au-Hg trimetallic amalgam with peroxidase-like activity. *Microchim. Acta* **2017**, *184*, 107–115. [CrossRef]
95. Li, W.; Bin, C.; Zhang, H.X.; Sun, Y.H.; Wang, J.; Zhang, J.L.; Fu, Y. BSA-stabilized Pt nanozyme for peroxidase mimetics and its application on colorimetric detection of mercury(II) ions. *Biosens. Bioelectron.* **2015**, *66*, 251–258. [CrossRef] [PubMed]

96. Zhou, Y.; Ma, Z.F. Fluorescent and colorimetric dual detection of mercury (II) by H_2O_2 oxidation of o-phenylenediamine using Pt nanoparticles as the catalyst. *Sens. Actuators B Chem.* **2017**, *249*, 53–58. [CrossRef]
97. Jiang, C.F.; Li, Z.J.; Wu, Y.X.; Guo, W.; Wang, J.S.; Jiang, Q. Colorimetric detection of Hg^{2+} based on enhancement of peroxidase-like activity of chitosan-gold nanoparticles. *Bull. Korean Chem. Soc.* **2018**, *39*, 625–630. [CrossRef]
98. Liao, H.; Liu, G.J.; Liu, Y.; Li, R.; Fu, W.S.; Hu, L.Z. Aggregation-induced accelerating peroxidase-like activity of gold nanoclusters and their applications for colorimetric Pb^{2+} detection. *Chem. Commun.* **2017**, *53*, 10160–10163. [CrossRef]
99. Zhang, W.C.; Niu, X.H.; Meng, S.C.; Li, X.; He, Y.F.; Pan, J.M.; Qiu, F.X.; Zhao, H.L.; Lan, M.B. Histidine-mediated tunable peroxidase-like activity of nanosized Pd for photometric sensing of Ag^+. *Sens. Actuators B Chem.* **2018**, *273*, 400–407. [CrossRef]
100. Chang, Y.; Zhang, Z.; Hao, J.; Yang, W.; Tang, J. BSA-stabilized Au clusters as peroxidase mimetic for colorimetric detection of Ag^+. *Sens. Actuators B Chem.* **2016**, *232*, 692–697. [CrossRef]
101. Long, Y.J.; Li, Y.F.; Liu, Y.; Zheng, J.J.; Tang, J.; Huang, C.Z. Visual observation of the mercury-stimulated peroxidase mimetic activity of gold nanoparticles. *Chem. Commun.* **2011**, *47*, 11939–11941. [CrossRef]
102. Niu, X.H.; He, Y.F.; Li, X.; Zhao, H.L.; Pan, J.M.; Qiu, F.X.; Lan, M.B. A peroxidase-mimicking nanosensor with Hg^{2+}-triggered enzymatic activity of cysteine-decorated ferromagnetic particles for ultrasensitive Hg^{2+} detection in environmental and biological fluids. *Sens. Actuators B Chem.* **2019**, *281*, 445–452. [CrossRef]
103. Huang, L.J.; Zhu, Q.R.; Zhu, J.; Luo, L.P.; Pu, S.H.; Zhang, W.T.; Zhu, W.X.; Sun, J.; Wang, J.L. Portable colorimetric detection of mercury(II) based on a non-noble metal nanozyme with tunable activity. *Inorg. Chem.* **2019**, *58*, 1638–1646. [CrossRef] [PubMed]
104. Mu, J.S.; Li, J.; Zhao, X.; Yang, E.C.; Zhao, X.J. Novel urchin-like Co_9S_8 nanomaterials with efficient intrinsic peroxidase-like activity for colorimetric sensing of copper (II) ion. *Sens. Actuators B Chem.* **2018**, *258*, 32–41. [CrossRef]
105. Zhang, Y.M.; Song, J.; Pan, Q.L.; Zhang, X.; Shao, W.H.; Zhang, X.; Quan, C.S.; Li, J. An Au@NH_2-MIL-125(Ti)-based multifunctional platform for colorimetric detections of biomolecules and Hg^{2+}. *J. Mater. Chem. B* **2020**, *8*, 114–124. [CrossRef] [PubMed]
106. Li, C.R.; Hai, J.; Fan, L.; Li, S.L.; Wang, B.D.; Yang, Z.Y. Amplified colorimetric detection of Ag^+ based on Ag^+-triggered peroxidase-like catalytic activity of ZIF-8/GO nanosheets. *Sens. Actuators B Chem.* **2019**, *284*, 213–219. [CrossRef]
107. Wang, Y.; Liang, R.P.; Qiu, J.D. Nanoceria-templated metal organic frameworks with oxidase-mimicking activity boosted by hexavalent chromium. *Anal. Chem.* **2020**, *92*, 2339–2346. [CrossRef]
108. Ma, C.; Ma, Y.; Sun, Y.; Lu, Y.; Tian, E.; Lan, J.; Li, J.; Ye, W.; Zhang, H. Colorimetric determination of Hg^{2+} in environmental water based on the Hg^{2+}-stimulated peroxidase mimetic activity of MoS_2-Au composites. *J. Colloid. Interfaces Sci.* **2019**, *537*, 554–561. [CrossRef]
109. Zhang, S.; Li, H.; Wang, Z.; Liu, J.; Zhang, H.; Wang, B.; Yang, Z. A strongly coupled Au/Fe_3O_4/GO hybrid material with enhanced nanozyme activity for highly sensitive colorimetric detection, and rapid and efficient removal of Hg^{2+} in aqueous solutions. *Nanoscale* **2015**, *7*, 8495–8502. [CrossRef]
110. Guo, L.; Mao, L.; Huang, K.; Liu, H. Pt–Se nanostructures with oxidase-like activity and their application in a selective colorimetric assay for mercury(II). *J. Mater. Sci.* **2017**, *52*, 10738–10750. [CrossRef]
111. Cao, H.; Xiao, J.Y.; Liu, H.M. Enhanced oxidase-like activity of selenium nanoparticles stabilized by chitosan and application in a facile colorimetric assay for mercury (II). *Biochem. Eng. J.* **2019**, *152*, 107384. [CrossRef]
112. Han, L.; Zhang, H.J.; Li, F. Bioinspired nanozymes with pH-independent and metal ions-controllable activity: Field-programmable logic conversion of sole logic gate system. *Part. Part. Syst. Charact.* **2018**, *35*, 1800207. [CrossRef]
113. Liu, R.; Zuo, L.; Huang, X.R.; Liu, S.M.; Yang, G.Y.; Li, S.Y.; Lv, C.Y. Colorimetric determination of lead(II) or mercury(II) based on target induced switching of the enzyme-like activity of metallothionein-stabilized copper nanoclusters. *Microchim. Acta* **2019**, *186*, 250. [CrossRef] [PubMed]
114. Han, K.N.; Choi, J.S.; Kwon, J. Gold nanozyme-based paper chip for colorimetric detection of mercury ions. *Sci. Rep.* **2017**, *7*, 2806. [CrossRef] [PubMed]
115. Kora, A.J.; Rastogi, L. Peroxidase activity of biogenic platinum nanoparticles: A colorimetric probe towards selective detection of mercuric ions in water samples. *Sens. Actuators B Chem.* **2018**, *254*, 690–700. [CrossRef]

116. Wang, Y.; Irudayaraj, J. A SERS DNAzyme biosensor for lead ion detection. *Chem. Commun.* **2011**, *47*, 4394–4396. [CrossRef]
117. Xie, Z.J.; Shi, M.R.; Wang, L.Y.; Peng, C.F.; Wei, X.L. Colorimetric determination of Pb^{2+} ions based on surface leaching of Au@Pt nanoparticles as peroxidase mimic. *Microchim. Acta* **2020**, *187*, 255. [CrossRef]
118. Tang, Y.; Hu, Y.; Yang, Y.X.; Liu, B.Y.; Wu, Y.G. A facile colorimetric sensor for ultrasensitive and selective detection of lead(II) in environmental and biological samples based on intrinsic peroxidase-mimic activity of WS_2 nanosheets. *Anal. Chim. Acta* **2020**, *1106*, 115–125. [CrossRef]
119. Li, M.; Huang, X.; Yu, H. A colorimetric assay for ultrasensitive detection of copper (II) ions based on pH-dependent formation of heavily doped molybdenum oxide nanosheets. *Mater. Sci. Eng. C* **2019**, *101*, 614–618. [CrossRef]
120. Shen, Q.; Li, W.; Tang, S.; Hu, Y.; Nie, Z.; Huang, Y.; Yao, S. A simple "clickable" biosensor for colorimetric detection of copper(II) ions based on unmodified gold nanoparticles. *Biosens. Bioelectron.* **2013**, *41*, 663–668. [CrossRef]
121. Gao, M.; An, P.L.; Rao, H.H.; Niu, Z.R.; Xue, X.; Luo, M.Y.; Liu, X.H.; Xue, Z.H.; Lu, X.Q. Molecule-gated surface chemistry of Pt nanoparticles for constructing activity-controllable nanozymes and a three-in-one sensor. *Analyst* **2020**, *145*, 1279–1287. [CrossRef]
122. White, P.A.; Collis, G.E.; Skidmore, M.; Breedon, M.; Ganther, W.D.; Venkatesan, K. Towards materials discovery: Assays for screening and study of chemical interactions of novel corrosion inhibitors in solution and coatings. *New J. Chem.* **2020**, *44*, 7647–7658. [CrossRef]
123. World Health Organization—Arsenic Fact Sheet. Available online: https://www.who.int/news-room/fact-sheets/detail/arsenic (accessed on 6 December 2020).
124. Zhong, X.-L.; Wen, S.-H.; Wang, Y.; Luo, Y.-X.; Li, Z.-M.; Liang, R.-P.; Zhang, L.; Qiu, J.-D. Colorimetric and electrochemical arsenate assays by exploiting the peroxidase-like activity of FeOOH nanorods. *Microchim. Acta* **2019**, *186*, 732. [CrossRef] [PubMed]
125. Wen, S.-H.; Zhong, X.-L.; Wu, Y.-D.; Liang, R.-P.; Zhang, L.; Qiu, J.-D. Colorimetric assay conversion to highly sensitive electrochemical assay for bimodal detection of arsenate based on cobalt oxyhydroxide nanozyme via arsenate absorption. *Anal. Chem.* **2019**, *91*, 6487–6497. [CrossRef] [PubMed]
126. Peng, C.F.; Zhang, Y.Y.; Wang, L.Y.; Jin, Z.Y.; Shao, G. Colorimetric assay for the simultaneous detection of Hg^{2+} and Ag^+ based on inhibiting the peroxidase-like activity of core-shell Au@Pt nanoparticles. *Anal. Methods* **2017**, *9*, 4363–4370. [CrossRef]
127. Zhao, Y.; Yang, X.; Cui, L.Y.; Sun, Y.L.; Song, Q.J. PVP-capped Pt NPs-depended catalytic nanoprobe for the simultaneous detection of Hg^{2+} and Ag^+. *Dyes Pigment.* **2018**, *150*, 21–26. [CrossRef]
128. Aragay, G.; Pino, F.; Merkoci, A. Nanomaterials for sensing and destroying pesticides. *Chem. Rev.* **2012**, *112*, 5317–5338. [CrossRef]
129. Pope, C.N. Organophosphorus pesticides: Do they all have the same mechanism of toxicity? *J. Toxicol. Environ. Health B* **1999**, *2*, 161–181. [CrossRef]
130. Hernandez, F.; Sancho, J.V.; Pozo, O.J. Critical review of the application of liquid chromatography/mass spectrometry to the determination of pesticide residues in biological samples. *Anal. Bioanal. Chem.* **2005**, *382*, 934–946. [CrossRef]
131. Blesa, J.; Soriano, J.M.; Molto, J.C.; Marin, R.; Manes, J. Determination of aflatoxins in peanuts by matrix solid-phase dispersion and liquid chromatography. *J. Chromatogr. A* **2003**, *1011*, 49–54. [CrossRef]
132. Ellman, G.L.; Courtney, K.D.; Andres, V., Jr.; Feather-Stone, R.M. A new and rapid colorimetric determination of acetylcholinesterase activity. *Biochem. Pharmacol.* **1961**, *7*, 88–95. [CrossRef]
133. Maeda, H.; Matsuno, H.; Ushida, M.; Katayama, K.; Saeki, K.; Itoh, N. 2,4-Dinitrobenzenesulfonyl fluoresceins as fluorescent alternatives to Ellman's reagent in thiol-quantification enzyme assays. *Angew. Chem. Int. Ed.* **2005**, *44*, 2922–2925. [CrossRef] [PubMed]
134. Sabelle, S.; Renard, P.Y.; Pecorella, K.; de Suzzoni-Dezard, S.; Creminon, C.; Grassi, J.; Mioskowski, C. Design and synthesis of chemiluminescent probes for the detection of cholinesterase activity. *J. Am. Chem. Soc.* **2002**, *124*, 4874–4880. [CrossRef] [PubMed]
135. Wang, J.; Timchalk, C.; Lin, Y. Carbon nanotube-based electrochemical sensor for assay of salivary cholinesterase enzyme activity: An exposure biomarker of organophosphate pesticides and nerve agents. *Environ. Sci. Technol.* **2008**, *42*, 2688–2693. [CrossRef] [PubMed]

136. Scott, C.; Pandey, G.; Hartley, C.J.; Jackson, C.J.; Cheesman, M.J.; Taylor, M.C.; Pandey, R.; Khurana, J.L.; Teese, M.; Coppin, C.W.; et al. The enzymatic basis for pesticide bioremediation. *Indian J. Microbiol.* **2008**, *48*, 65–79. [CrossRef] [PubMed]
137. Sharma, B.; Dangi, A.K.; Shukla, P. Contemporary enzyme based technologies for bioremediation: A review. *J. Environ. Manag.* **2018**, *210*, 10–22. [CrossRef] [PubMed]
138. Singh, S.; Tripathi, P.; Kumar, N.; Nara, S. Colorimetric sensing of malathion using palladium-gold bimetallic nanozyme. *Biosens. Bioelectron.* **2017**, *92*, 280–286. [CrossRef] [PubMed]
139. Zhu, Y.Y.; Wu, J.J.X.; Han, L.J.; Wang, X.Y.; Li, W.; Guo, H.C.; Wei, H. Nanozyme sensor arrays based on heteroatom-doped graphene for detecting pesticides. *Anal. Chem.* **2020**, *92*, 7444–7452. [CrossRef]
140. Liang, M.; Fan, K.; Pan, Y.; Jiang, H.; Wang, F.; Yang, D.; Lu, D.; Feng, J.; Zhao, J.; Yang, L.; et al. Fe_3O_4 magnetic nanoparticle peroxidase mimetic-based colorimetric assay for the rapid detection of organophosphorus pesticide and nerve agent. *Anal. Chem.* **2013**, *85*, 308–312. [CrossRef]
141. Wei, J.C.; Yang, L.L.; Luo, M.; Wang, Y.T.; Li, P. Nanozyme-assisted technique for dual mode detection of organophosphorus pesticide. *Ecotoxicol. Environ. Saf.* **2019**, *179*, 17–23. [CrossRef]
142. Wei, J.; Yang, Y.; Dong, J.; Wang, S.; Li, P. Fluorometric determination of pesticides and organophosphates using nanoceria as a phosphatase mimic and an inner filter effect on carbon nanodots. *Microchim. Acta* **2019**, *186*, 66. [CrossRef]
143. Sun, Y.; Wei, J.; Zou, J.; Cheng, Z.; Huang, Z.; Gu, L.; Zhong, Z.; Li, S.; Wang, Y.; Li, P. Electrochemical detection of methyl-paraoxon based on bifunctional nanozyme with catalytic activity and signal amplification effect. *J. Pharm. Anal.* **2020**. [CrossRef]
144. Guan, G.; Yang, L.; Mei, Q.; Zhang, K.; Zhang, Z.; Han, M.Y. Chemiluminescence switching on peroxidase-like Fe_3O_4 nanoparticles for selective detection and simultaneous determination of various pesticides. *Anal. Chem.* **2012**, *84*, 9492–9497. [CrossRef] [PubMed]
145. Boruah, P.K.; Das, M.R. Dual responsive magnetic Fe_3O_4-TiO_2/graphene nanocomposite as an artificial nanozyme for the colorimetric detection and photodegradation of pesticide in an aqueous medium. *J. Hazard. Mater.* **2020**, *385*, 121516. [CrossRef] [PubMed]
146. Biswas, S.; Tripathi, P.; Kumar, N.; Nara, S. Gold nanorods as peroxidase mimetics and its application for colorimetric biosensing of malathion. *Sens. Actuators B Chem.* **2016**, *231*, 584–592. [CrossRef]
147. Weerathunge, P.; Behera, B.K.; Zihara, S.; Singh, M.; Prasad, S.N.; Hashmi, S.; Mariathomas, P.R.D.; Bansal, V.; Ramanathan, R. Dynamic interactions between peroxidase-mimic silver NanoZymes and chlorpyrifos-specific aptamers enable highly-specific pesticide sensing in river water. *Anal. Chim. Acta* **2019**, *1083*, 157–165. [CrossRef]
148. Weerathunge, P.; Ramanathan, R.; Shukla, R.; Sharma, T.K.; Bansal, V. Aptamer-controlled reversible inhibition of gold nanozyme activity for pesticide sensing. *Anal. Chem.* **2014**, *86*, 11937–11941. [CrossRef]
149. Xia, W.Q.; Zhang, P.; Fu, W.S.; Hu, L.Z.; Wang, Y. Aggregation/dispersion-mediated peroxidase-like activity of MoS_2 quantum dots for colorimetric pyrophosphate detection. *Chem. Commun.* **2019**, *55*, 2039–2042. [CrossRef]
150. Kushwaha, A.; Singh, G.; Sharma, M. Colorimetric sensing of chlorpyrifos through negative feedback inhibition of the catalytic activity of silver phosphate oxygenase nanozymes. *RSC Adv.* **2020**, *10*, 13050–13065. [CrossRef]
151. Wijaya, W.; Pang, S.; Labuza, T.P.; He, L. Rapid detection of acetamiprid in foods using surface-enhanced Raman spectroscopy (SERS). *J. Food Sci.* **2014**, *79*, T743–T747. [CrossRef]
152. Barrios-Estrada, C.; de Jesus Rostro-Alanis, M.; Munoz-Gutierrez, B.D.; Iqbal, H.M.N.; Kannan, S.; Parra-Saldivar, R. Emergent contaminants: Endocrine disruptors and their laccase-assisted degradation—A review. *Sci. Total Environ.* **2018**, *612*, 1516–1531. [CrossRef] [PubMed]
153. Cai, S.; Han, Q.; Qi, C.; Wang, X.; Wang, T.; Jia, X.; Yang, R.; Wang, C. MoS_2-Pt_3Au_1 nanocomposites with enhanced peroxidase-like activities for selective colorimetric detection of phenol. *Chin. J. Chem.* **2017**, *35*, 605–612. [CrossRef]
154. Zhang, J.; Zhuang, J.; Gao, L.; Zhang, Y.; Gu, N.; Feng, J.; Yang, D.; Zhu, J.; Yan, X. Decomposing phenol by the hidden talent of ferromagnetic nanoparticles. *Chemosphere* **2008**, *73*, 1524–1528. [CrossRef] [PubMed]
155. Liu, Y.; Zhu, G.; Bao, C.; Yuan, A.; Shen, X. Intrinsic peroxidase-like activity of porous CuO micro-/nanostructures with clean surface. *Chin. J. Chem.* **2014**, *32*, 151–156. [CrossRef]

156. Feng, Y.B.; Hong, L.; Liu, A.L.; Chen, W.D.; Li, G.W.; Chen, W.; Xia, X.H. High-efficiency catalytic degradation of phenol based on the peroxidase-like activity of cupric oxide nanoparticles. *Int. J. Environ. Sci. Technol.* **2015**, *12*, 653–660. [CrossRef]
157. Jiang, J.; He, C.; Wang, S.; Jiang, H.; Li, J.; Li, L. Recyclable ferromagnetic chitosan nanozyme for decomposing phenol. *Carbohyd. Polym.* **2018**, *198*, 348–353. [CrossRef]
158. Xu, L.; Wang, J. Magnetic nanoscaled Fe_3O_4/CeO_2 composite as an efficient Fenton-like heterogeneous catalyst for degradation of 4-chlorophenol. *Environ. Sci. Technol.* **2012**, *46*, 10145–10153. [CrossRef]
159. Sadaf, A.; Ahmad, R.; Ghorbal, A.; Elfalleh, W.; Khare, S.K. Synthesis of cost-effective magnetic nano-biocomposites mimicking peroxidase activity for remediation of dyes. *Environ. Sci. Pollut. Res. Int.* **2019**, *27*, 27211–27220. [CrossRef]
160. Bhuyan, D.; Arbuj, S.S.; Saikia, L. Template-free synthesis of Fe_3O_4 nanorod bundles and their highly efficient peroxidase mimetic activity for the degradation of organic dye pollutants with H_2O_2. *New J. Chem.* **2015**, *39*, 7759–7762. [CrossRef]
161. Pariona, N.; Herrera-Trejo, M.; Oliva, J.; Martinez, A.I. Peroxidase-like activity of ferrihydrite and hematite nanoparticles for the degradation of methylene blue. *J. Nanomater.* **2016**, *2016*, 1–8. [CrossRef]
162. Ma, J.; Zhai, G. Antibiotic contamination: A global environment issue. *J. Bioremed. Biodegrad.* **2014**, *5*, e157. [CrossRef]
163. Shen, Y.; Zhao, W.; Zhang, C.; Shan, Y.; Shi, J. Degradation of streptomycin in aquatic environment: Kinetics, pathway, and antibacterial activity analysis. *Environ. Sci. Pollut. Res. Int.* **2017**, *24*, 14337–14345. [CrossRef] [PubMed]
164. Zhao, J.; Wu, Y.; Tao, H.; Chen, H.; Yang, W.; Qiu, S. Colorimetric detection of streptomycin in milk based on peroxidase-mimicking catalytic activity of gold nanoparticles. *RSC Adv.* **2017**, *7*, 38471–38478. [CrossRef]
165. Sharma, T.K.; Ramanathan, R.; Weerathunge, P.; Mohammadtaheri, M.; Daima, H.K.; Shukla, R.; Bansal, V. Aptamer-mediated 'turn-off/turn-on' nanozyme activity of gold nanoparticles for kanamycin detection. *Chem. Commun.* **2014**, *50*, 15856–15859. [CrossRef] [PubMed]
166. Hussain, C.M. *The Handbook of Environmental Remediation: Classic and Modern Techniques*; Royal Society of Chemistry: Cambridge, UK, 2020.
167. Wang, S.; Fang, H.; Yi, X.; Xu, Z.; Xie, X.; Tang, Q.; Ou, M.; Xu, X. Oxidative removal of phenol by HRP-immobilized beads and its environmental toxicology assessment. *Ecotoxicol. Environ. Saf.* **2016**, *130*, 234–239. [CrossRef]
168. He, J.; Liang, M. Nanozymes for environmental monitoring and treatment. In *Nanozymology*; Yan, X., Ed.; Springer: Berlin, Germany, 2020; pp. 527–543.
169. Harrad, S. *Persistent Organic Pollutants*; Wiley: Chichester, UK, 2010.
170. Cheng, R.; Li, G.-q.; Cheng, C.; Shi, L.; Zheng, X.; Ma, Z. Catalytic oxidation of 4-chlorophenol with magnetic Fe_3O_4 nanoparticles: Mechanisms and particle transformation. *RSC Adv.* **2015**, *5*, 66927–66933. [CrossRef]
171. Ulson de Souza, S.M.A.G.; Forgiarini, E.; Ulson de Souza, A.A. Toxicity of textile dyes and their degradation by the enzyme horseradish peroxidase (HRP). *J. Hazard. Mater.* **2007**, *147*, 1073–1078. [CrossRef]
172. Janoš, P.; Kuráň, P.; Pilařová, V.; Trögl, J.; Šťastný, M.; Pelant, O.; Henych, J.; Bakardjieva, S.; Životský, O.; Kormunda, M.; et al. Magnetically separable reactive sorbent based on the $CeO_2/\gamma\text{-}Fe_2O_3$ composite and its utilization for rapid degradation of the organophosphate pesticide parathion methyl and certain nerve agents. *Chem. Eng. J.* **2015**, *262*, 747–755. [CrossRef]
173. Wang, S.; Bromberg, L.; Schreuder-Gibson, H.; Hatton, T.A. Organophophorous ester degradation by chromium(III) terephthalate metal–organic framework (MIL-101) chelated to N,N-dimethylaminopyridine and related aminopyridines. *ACS Appl. Mater. Interfaces* **2013**, *5*, 1269–1278. [CrossRef]
174. Wang, X.; Liu, J.; Qu, R.; Wang, Z.; Huang, Q. The laccase-like reactivity of manganese oxide nanomaterials for pollutant conversion: Rate analysis and cyclic voltammetry. *Sci. Rep.* **2017**, *7*, 7756. [CrossRef]
175. Wang, J.; Huang, R.; Qi, W.; Su, R.; Binks, B.P.; He, Z. Construction of a bioinspired laccase-mimicking nanozyme for the degradation and detection of phenolic pollutants. *Appl. Catal. B Environ.* **2019**, *254*, 452–462. [CrossRef]
176. Huang, R.; Fang, Z.; Yan, X.; Cheng, W. Heterogeneous sono-Fenton catalytic degradation of bisphenol A by Fe_3O_4 magnetic nanoparticles under neutral condition. *Chem. Eng. J.* **2012**, *197*, 242–249. [CrossRef]

177. Wang, N.; Zhu, L.; Wang, D.; Wang, M.; Lin, Z.; Tang, H. Sono-assisted preparation of highly-efficient peroxidase-like Fe_3O_4 magnetic nanoparticles for catalytic removal of organic pollutants with H_2O_2. *Ultrason. Sonochem.* **2010**, *17*, 526–533. [CrossRef] [PubMed]
178. Yan, J.; Lei, M.; Zhu, L.; Anjum, M.N.; Zou, J.; Tang, H. Degradation of sulfamonomethoxine with Fe_3O_4 magnetic nanoparticles as heterogeneous activator of persulfate. *J. Hazard. Mater.* **2011**, *186*, 1398–1404. [CrossRef] [PubMed]
179. Niu, H.; Zhang, D.; Zhang, S.; Zhang, X.; Meng, Z.; Cai, Y. Humic acid coated Fe_3O_4 magnetic nanoparticles as highly efficient Fenton-like catalyst for complete mineralization of sulfathiazole. *J. Hazard. Mater.* **2011**, *190*, 559–565. [CrossRef] [PubMed]
180. Xu, L.; Wang, J. Fenton-like degradation of 2,4-dichlorophenol using Fe_3O_4 magnetic nanoparticles. *Appl. Catal. B Environ.* **2012**, *123–124*, 117–126. [CrossRef]
181. Zeb, A.; Xie, X.; Yousaf, A.B.; Imran, M.; Wen, T.; Wang, Z.; Guo, H.-L.; Jiang, Y.-F.; Qazi, I.A.; Xu, A.-W. Highly efficient Fenton and enzyme-mimetic activities of mixed-phase VO_x nanoflakes. *ACS Appl. Mater. Interfaces* **2016**, *8*, 30126–30132. [CrossRef] [PubMed]
182. Jain, S.; Panigrahi, A.; Sarma, T.K. Counter anion-directed growth of iron oxide nanorods in a polyol medium with efficient peroxidase-mimicking activity for degradation of dyes in contaminated water. *ACS Omega* **2019**, *4*, 13153–13164. [CrossRef]
183. Wang, Y.; Liu, T.; Liu, J. Synergistically boosted degradation of organic dyes by CeO_2 nanoparticles with fluoride at low pH. *ACS Appl. Nano Mater.* **2020**, *3*, 842–849. [CrossRef]
184. Safavi, A.; Sedaghati, F.; Shahbaazi, H.; Farjami, E. Facile approach to the synthesis of carbon nanodots and their peroxidase mimetic function in azo dyes degradation. *RSC Adv.* **2012**, *2*, 7367–7370. [CrossRef]
185. Xu, L.; Wang, J. A heterogeneous Fenton-like system with nanoparticulate zero-valent iron for removal of 4-chloro-3-methyl phenol. *J. Hazard. Mater.* **2011**, *186*, 256–264. [CrossRef]
186. Chen, T.M.; Xiao, J.; Wang, G.W. Cubic boron nitride with an intrinsic peroxidase-like activity. *RSC Adv.* **2016**, *6*, 70124–70132. [CrossRef]
187. Zhang, P.; Sun, D.; Cho, A.; Weon, S.; Lee, S.; Lee, J.; Han, J.W.; Kim, D.-P.; Choi, W. Modified carbon nitride nanozyme as bifunctional glucose oxidase-peroxidase for metal-free bioinspired cascade photocatalysis. *Nat. Commun.* **2019**, *10*, 940. [CrossRef] [PubMed]
188. Zuo, X.; Peng, C.; Huang, Q.; Song, S.; Wang, L.; Li, D.; Fan, C. Design of a carbon nanotube/magnetic nanoparticle-based peroxidase-like nanocomplex and its application for highly efficient catalytic oxidation of phenols. *Nano Res.* **2009**, *2*, 617–623. [CrossRef]
189. Peng, C.; Jiang, B.; Liu, Q.; Guo, Z.; Xu, Z.; Huang, Q.; Xu, H.; Tai, R.; Fan, C. Graphene-templated formation of two-dimensional lepidocrocite nanostructures for high-efficiency catalytic degradation of phenols. *Energy Environ. Sci.* **2011**, *4*, 2035–2040. [CrossRef]
190. Chun, J.; Lee, H.; Lee, S.-H.; Hong, S.-W.; Lee, J.; Lee, C.; Lee, J. Magnetite/mesocellular carbon foam as a magnetically recoverable fenton catalyst for removal of phenol and arsenic. *Chemosphere* **2012**, *89*, 1230–1237. [CrossRef]
191. Ding, Y.; Li, Z.; Jiang, W.; Yuan, B.; Huang, T.; Wang, L.; Tang, J. Degradation of phenol using a peroxidase mimetic catalyst through conjugating deuterohemin-peptide onto metal-organic framework with enhanced catalytic activity. *Catal. Commun.* **2020**, *134*, 105859. [CrossRef]
192. Tian, S.H.; Tu, Y.T.; Chen, D.S.; Chen, X.; Xiong, Y. Degradation of acid Orange II at neutral pH using $Fe_2(MoO_4)_3$ as a heterogeneous Fenton-like catalyst. *Chem. Eng. J.* **2011**, *169*, 31–37. [CrossRef]
193. Zhang, Z.; Hao, J.; Yang, W.; Lu, B.; Ke, X.; Zhang, B.; Tang, J. Porous Co_3O_4 nanorods–reduced graphene oxide with intrinsic peroxidase-like activity and catalysis in the degradation of methylene blue. *ACS Appl. Mater. Interfaces* **2013**, *5*, 3809–3815. [CrossRef]
194. Zubir, N.A.; Yacou, C.; Motuzas, J.; Zhang, X.; Diniz da Costa, J.C. Structural and functional investigation of graphene oxide–Fe_3O_4 nanocomposites for the heterogeneous Fenton-like reaction. *Sci. Rep.* **2014**, *4*, 4594. [CrossRef]
195. Hu, P.; Han, L.; Dong, S. A facile one-pot method to synthesize a polypyrrole/hemin nanocomposite and its application in biosensor, dye removal, and photothermal therapy. *ACS Appl. Mater. Interfaces* **2014**, *6*, 500–506. [CrossRef]
196. Wan, D.; Li, W.; Wang, G.; Chen, K.; Lu, L.; Hu, Q. Adsorption and heterogeneous degradation of rhodamine B on the surface of magnetic bentonite material. *Appl. Surf. Sci.* **2015**, *349*, 988–996. [CrossRef]

197. Deng, J.; Wen, X.; Li, J. Fabrication highly dispersed Fe$_3$O$_4$ nanoparticles on carbon nanotubes and its application as a mimetic enzyme to degrade Orange II. *Environ. Technol.* **2016**, *37*, 2214–2221. [CrossRef] [PubMed]
198. Li, S.; Hou, Y.; Chen, Q.; Zhang, X.; Cao, H.; Huang, Y. Promoting active sites in MOF-derived homobimetallic hollow nanocages as a high-performance multifunctional nanozyme catalyst for biosensing and organic pollutant degradation. *ACS Appl. Mater. Interfaces* **2019**, *12*, 2581–2590. [CrossRef] [PubMed]
199. Variava, M.F.; Church, T.L.; Harris, A.T. Magnetically recoverable Fe$_x$O$_y$–MWNT Fenton's catalysts that show enhanced activity at neutral pH. *Appl. Catal. B Environ.* **2012**, *123*, 200–207. [CrossRef]
200. Liu, W.; Qian, J.; Wang, K.; Xu, H.; Jiang, D.; Liu, Q.; Yang, X.; Li, H. Magnetically separable Fe$_3$O$_4$ nanoparticles-decorated reduced graphene oxide nanocomposite for catalytic wet hydrogen peroxide oxidation. *J. Inorg. Organomet. Polym. Mater.* **2013**, *23*, 907–916. [CrossRef]
201. Chang, Y.H.; Yao, Y.F.; Luo, H.; Cui, L.; Zhi, L.J. Magnetic Fe$_3$O$_4$-GO nanocomposites as highly efficient Fenton-like catalyst for the degradation of dyes. *Int. J. Nanomanuf.* **2014**, *10*, 132–141. [CrossRef]
202. Ribeiro, R.S.; Silva, A.M.T.; Figueiredo, J.L.; Faria, J.L.; Gomes, H.T. Catalytic wet peroxide oxidation: A route towards the application of hybrid magnetic carbon nanocomposites for the degradation of organic pollutants. A review. *Appl. Catal. B Environ.* **2016**, *187*, 428–460. [CrossRef]
203. Niu, H.; Meng, Z.; Cai, Y. Fast defluorination and removal of norfloxacin by alginate/Fe@Fe$_3$O$_4$ core/shell structured nanoparticles. *J. Hazard. Mater.* **2012**, *227-228*, 195–203. [CrossRef]
204. Zhu, M.; Diao, G. Synthesis of porous Fe$_3$O$_4$ nanospheres and its application for the catalytic degradation of xylenol orange. *J. Phys. Chem. C* **2011**, *115*, 18923–18934. [CrossRef]
205. Wang, L.; Zeng, Y.; Shen, A.; Zhou, X.; Hu, J. Three dimensional nano-assemblies of noble metal nanoparticle-infinite coordination polymers as specific oxidase mimetics for degradation of methylene blue without adding any cosubstrate. *Chem. Commun.* **2015**, *51*, 2052–2055. [CrossRef] [PubMed]
206. Zhao, C.; Xiong, C.; Liu, X.; Qiao, M.; Li, Z.; Yuan, T.; Wang, J.; Qu, Y.; Wang, X.; Zhou, F.; et al. Unraveling the enzyme-like activity of heterogeneous single atom catalyst. *Chem. Commun.* **2019**, *55*, 2285–2288. [CrossRef]
207. Kucharzyk, K.H.; Darlington, R.; Benotti, M.; Deeb, R.; Hawley, E. Novel treatment technologies for PFAS compounds: A critical review. *J. Environ. Manag.* **2017**, *204*, 757–764. [CrossRef]
208. Ross, I.; McDonough, J.; Miles, J.; Storch, P.; Thelakkat Kochunarayanan, P.; Kalve, E.; Hurst, J.; Dasgupta, S.S.; Burdick, J. A review of emerging technologies for remediation of PFASs. *Remediation* **2018**, *28*, 101–126. [CrossRef]
209. Liu, J.; Avendaño, S.M. Microbial degradation of polyfluoroalkyl chemicals in the environment: A review. *Environ. Int.* **2013**, *61*, 98–114. [CrossRef]
210. Colosi, L.M.; Pinto, R.A.; Huang, Q.; Weber, W.J., Jr. Peroxidase-mediated degradation of perfluorooctanoic acid. *Environ. Toxicol. Chem.* **2009**, *28*, 264–271. [CrossRef]
211. Luo, Q.; Lu, J.; Zhang, H.; Wang, Z.; Feng, M.; Chiang, S.-Y.D.; Woodward, D.; Huang, Q. Laccase-catalyzed degradation of perfluorooctanoic acid. *Environ. Sci. Technol. Lett.* **2015**, *2*, 198–203. [CrossRef]

Publisher's Note: MDPI stays neutral with regard to jurisdictional claims in published maps and institutional affiliations.

© 2021 by the authors. Licensee MDPI, Basel, Switzerland. This article is an open access article distributed under the terms and conditions of the Creative Commons Attribution (CC BY) license (http://creativecommons.org/licenses/by/4.0/).

Review

Resonance Energy Transfer-Based Biosensors for Point-of-Need Diagnosis—Progress and Perspectives

Felix Weihs [1], Alisha Anderson [2], Stephen Trowell [3] and Karine Caron [2,*]

[1] CSIRO Health & Biosecurity, Parkville, 343 Royal Parade, Melbourne, VIC 3030, Australia; felix.weihs@csiro.au
[2] CSIRO Health & Biosecurity, Black Mountain, Canberra, ACT 2600, Australia; alisha.anderson@csiro.au
[3] PPB Technology Pty Ltd., Centre for Entrepreneurial Agri-Technology, Australian National University, Canberra, ACT 2601, Australia; sct.ppbtech@icloud.com
* Correspondence: karine.caron@csiro.au; Tel.: +61-2626-464-152

Received: 17 December 2020; Accepted: 15 January 2021; Published: 19 January 2021

Abstract: The demand for point-of-need (PON) diagnostics for clinical and other applications is continuing to grow. Much of this demand is currently serviced by biosensors, which combine a bioanalytical sensing element with a transducing device that reports results to the user. Ideally, such devices are easy to use and do not require special skills of the end user. Application-dependent, PON devices may need to be capable of measuring low levels of analytes very rapidly, and it is often helpful if they are also portable. To date, only two transduction modalities, colorimetric lateral flow immunoassays (LFIs) and electrochemical assays, fully meet these requirements and have been widely adopted at the point-of-need. These modalities are either non-quantitative (LFIs) or highly analyte-specific (electrochemical glucose meters), therefore requiring considerable modification if they are to be co-opted for measuring other biomarkers. Förster Resonance Energy Transfer (RET)-based biosensors incorporate a quantitative and highly versatile transduction modality that has been extensively used in biomedical research laboratories. RET-biosensors have not yet been applied at the point-of-need despite its advantages over other established techniques. In this review, we explore and discuss recent developments in the translation of RET-biosensors for PON diagnoses, including their potential benefits and drawbacks.

Keywords: FRET; BRET; CRET; point-of-care; on-site; on-the-spot; microfluidics; PADs; time-resolved FRET

1. Introduction

Over the last 50 years, there has been a substantial trend to developing point-of-need (PON) diagnostic testing, also known as "on-the-spot" or "point-of-care", putting rapid testing into the hands of first responders, processors and consumers. The development of such diagnostics is seen across healthcare [1,2], food and beverage [3,4], and industrial quality control applications [5,6]. Ideally, a PON test delivers accurate and repeatable results in short time periods, with minimal sample preparation, minimal user expertise and minimal resource requirements. PON testing can be enabled by the miniaturization of a range of technologies such as ultrasound [7], MRI [8] and spectroscopy [9]. However, imaging and spectroscopic technologies tend to be limited to anatomically apparent pathologies and require significant on-site analysis and end-user expertise to interpret results. Biosensors that combine a biological recognition element with a transduction modality for the detection of a range of analytes are ideally placed to meet the requirements for accurate and repeatable PON diagnostics without deep analytical expertise on the part of the end-user. The biological recognition elements selected are proteins or nucleic acids that have been honed by natural selection to be highly sensitive and specific for their target analytes. The range of such elements includes enzymes, antibodies, receptor proteins, DNA and RNA. Biological recognition elements can be coupled to a mechanical, electrical or optical transduction mechanism, to enable rapid signal readouts, and many different biosensor transduction mechanisms have been used in research and diagnostic laboratories. For example, fluorescence [10], Surface Plasmon Resonance (SPR) [11] and Surface Enhanced Raman Spectroscopy (SERS) [12] have recently attracted significant interest in research for PON biosensing. However, only two transduction mechanisms have been widely adopted at the PON, namely lateral flow immunoassay (LFI) and electrochemical (EC) readouts.

LFI is widely used in many fields, including medicine [13], veterinary [14], food [4], agricultural [15] and industrial process control [16]. As an example, the most commonly used LFI is the pregnancy test, which is available on many pharmacy shelves. The principle of LFIs is based on the movement of a liquid sample via capillary action through a paper-based strip containing antibodies with which analytes of interest interact to produce a colorimetric readout [17]. While LFIs are simple, cheap and portable, the results obtained are qualitative or at best semi-quantitative [18], limiting them to use as a screening tool. Additionally, many LFIs are not sensitive enough to measure trace amounts of analytes in complex samples [19], and they are sometimes unreliable due to the limited control of sample volumes and user error [20,21].

Electrochemical transduction, as the name suggests, utilizes outputs of a voltage or current when biological-recognition elements immobilized on an electrode interact with the target analyte. The best-known application of electrochemical biosensors is for the quantification of glucose for diabetes management [22].

Electrochemical biosensors offer the advantage of simplicity, low cost and reliable quantitative detection even with complex sample matrices. However, they tend to be challenged when detecting trace amounts of analytes and suffer from biosensor drift and surface effects on the electrodes after repeated exposure to biological or chemical matrices, further reducing the sensitivity [23,24]. Very recently, as a means to respond to the worldwide healthcare crisis, rapid detection of SARS-CoV-2 has been demonstrated [10]. With the aim of moving rapid tests from laboratory settings to the PON, CRISPR-based technologies are now being adapted to give either an electrochemical readout [25] or to be carried out on a paper strip [10] similar to that of LFIs.

In this study, we investigated whether there are opportunities for emerging transduction modalities to bring new diagnostic tests to the point-of-care. Specifically, we focused on a class of luminometry, namely Förster Resonance Energy Transfer (RET), that was elaborately developed for research and laboratory use. We explored to what extent, and under what conditions, RET might open up new opportunities in biosensing at the point-of-need.

2. Förster Resonance Energy Transfer Sensing Principle

RET is a distance- and orientation-dependent non-radiative energy transfer from an energy donor to an acceptor fluorophore or quencher [26]. Resonance Energy Transfer usually occurs at distances of 1–10 nm, which can vary dependent on the combination of donor and acceptor and the relative orientation of their transition dipole moments [27–29]. Due to its extreme sensitivity to spatial changes on a nanometer scale, RET has been extensively used as the transduction modality in lab-based biosensors for the analysis of a wide range of analyte types, such as kinase [30] and protease activity [31–33], G-protein-coupled receptors [34,35], antibodies [36], small molecules [37,38] and protein–protein interaction [39–41].

Biosensors have been realized through the incorporation of RET components into biological recognition elements in such a way that the presence of an analyte affects the spatial relation between the donor and acceptor molecules (Figure 1). The versatility of this approach facilitates theoretically infinite options for sensing applications. In contrast, glucose meters rely on a specific enzymatic reaction causing the oxidation of glucose [42]. Such a mechanism cannot be readily translated for the detection of other analytes.

We can identify three different classes of RET according to the identity of the incorporated energy donor: (1) Fluorescence Resonance Energy Transfer (FRET), (2) Bioluminescence Resonance Energy Transfer (BRET) and Chemiluminescence Resonance Energy Transfer (CRET) (Figure 1a). FRET is, arguably, the most frequently applied RET technique in lab-based testing, where energy transfer is initiated through the excitation of a fluorescent molecule, such as a fluorescent protein, organic dye, quantum dot or rare earth element, using an external illumination source. Unlike FRET, BRET and CRET employ endogenous "light" sources by the incorporation of enzymes or chemical catalysts as the energy donor. They require suitable chemical substrates to generate luminescence. Formally speaking, BRET is a subclass of CRET but will be discussed separately because it has several distinct features. The selection of energy donor has profound implications for the technical requirements testing outside the research laboratory, by determining how the optical response of the RET biosensors is initiated. External illumination can generate significant noise due to light scattering, autofluorescence, and bleed through and crosstalk of the exciting radiation. BRET/CRET can therefore deliver reduced noise levels, particularly in complex samples such as blood [43]. Since noise is often limiting for sensitivity, BRET/CRET is potentially well-suited for PON applications, enabling more reproducible measurements and the use of smaller sample amounts. As no external light source component is required, the design of miniaturized BRET/CRET detection devices is simplified compared to in the use of FRET biosensors.

Luciferases have been applied in biolanalytics for decades, but, until recently, bioluminescence and BRET transduction techniques were not considered for point-of-need applications. Their low optical signal output was particularly challenging for miniaturized applications, where low amounts of biosensors are used. However, considerable progress has been made in recent years in the development of brighter and more stable luciferase-luciferin combinations [44,45] coupled to superior acceptors [46,47] and in combination with sensitive BRET detection devices [48].

Enzyme degradation, substrate instability or substrate inhibition can affect the amount of light produced by the luciferase oxidation reaction. However, the variation of emitted photons can be alleviated by using excess substrate and by using ratiometric measurements, as, to a certain extent, a lower light generation is not accompanied by a change in signal ratio. However, some luciferases exhibit flash-type kinetics, demanding a timed detection system, which can be a drawback compared to using FRET as the signal transduction modality.

3. Signal Measurements

RET signals within the context of on-site suitable techniques are usually measured either by sensitized emission or fluorescence lifetime. Sensitized emission is commonly applied for all RET systems and describes the measurement of donor and acceptor signal intensities in relation to each other (RET ratio). Close proximity between donor and acceptor molecules results in higher RET ratios than more distant RET components.

Ratiometric measurements exhibit a dose-independent normalization, as the RET ratios for 10 molecules are theoretically the same as for 100 molecules. This makes RET techniques more robust to interfering effects, compared to fluorometric or luminometric approaches, where only the signal of a single fluorophore/luminophore is measured. RET therefore enables homogeneous assays without the need for washing steps. Another way of achieving internal normalization is to analyze the fluorescence lifetime of the donor (time-resolved FRET). The time between donor excitation and photon emission is prolonged if there is RET to an acceptor molecule. Fluorescence lifetimes are usually in the nano- to micro-second range, requiring sophisticated analytical equipment. Fluorescence lifetime is the method of choice for Lanthanide RET (LRET) systems, where rare earth elements with remarkably long fluorescence lifetimes are incorporated as energy donors [49,50].

Whether it is RET ratio or fluorescence lifetime that is measured, these parameters are usually calibrated against standard concentrations of the target analyte, to generate results that are meaningful to the end-user.

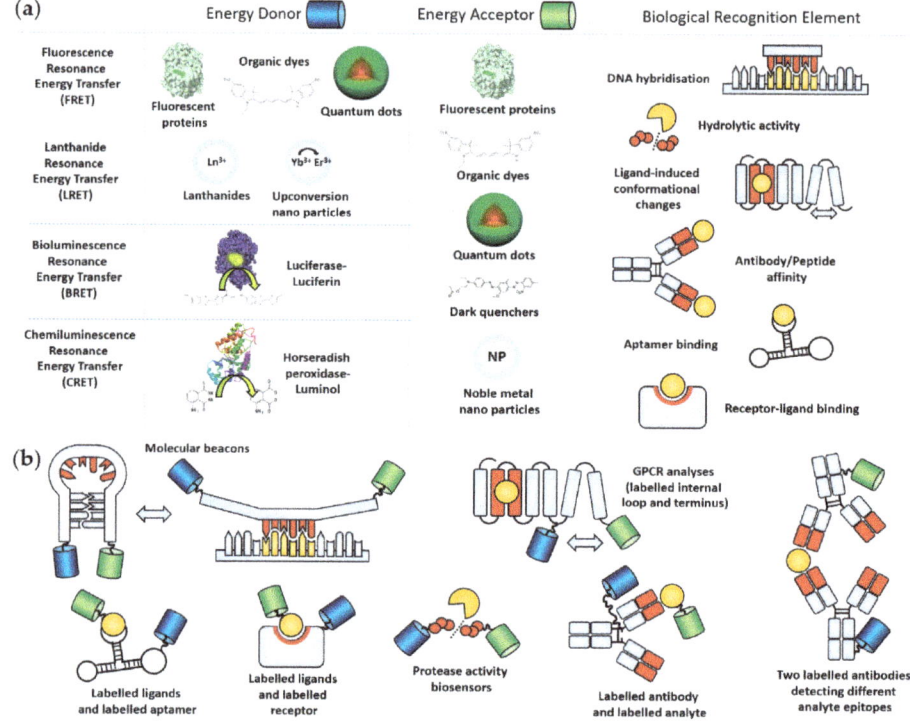

Figure 1. (**a**) Overview of different variations on the Resonance Energy Transfer principle, together with a range of different types of biological recognition element. Red domains indicate recognition elements, and orange domains indicate targeted analytes. (**b**) Examples of different ways that energy donors and acceptors can be combined with biological recognition elements. All illustrations are simplified and not shown to scale. Images were taken from the following sources: The "Fluorescent proteins" image illustrates the Green Fluorescent Protein (doi:10.2210/rcsb_pdb/mom_2003_6). The "Organic dyes" structure shows cyanine. The quantum dot image was taken from Reference [49]. The "Luciferase–Luciferin" image is composed of the Firefly luciferase (doi:10.2210/rcsb_pdb/mom_2006_6) and D-Luciferin as its substrate. The structure of horseradish peroxidase was taken from the Protein Data Bank (1W4E, doi:10.2210/pdb1W4W/pdb). The dark quencher is the black hole quencher BHQ1 from atdbio (https://www.atdbio.com/content/35/FRET-fluorescence-quenchers).

4. PON-Suitable RET Applications

In a laboratory environment, RET biosensors are usually deployed in micro well plates and read out in a sophisticated plate-reading instrument. Such instruments tend to be too expensive and cumbersome for PON use. Alternative instruments have been developed that generally feature two simplifications. Firstly, micro titer plates are replaced by a medium, such as microfluidic channels, paper-strips or cartridges, that makes it easier to analyze small sample volumes with high reproducibility. Secondly, a PON-specific device is used to detect and interpret RET signals. Approaches include the use of compact integrated devices, modified microscopes or digital cameras.

It is noteworthy that the overall sensitivity of a biosensing test, either lab-based or at the PON, depends on the sensitivity of the biological element used, the transduction modality and the equipment used. In the case of PON, it is possible that sensitivity or quantification may be reduced as a tradeoff for using more convenient equipment. Consequently, the sensitivity of a test depends on all of the components involved. As an indication of the performance of each type of technology, we have listed approximate assay times and sensitivity levels in the respective figures summarizing the technologies reviewed in this article.

4.1. RET Detection Using Microfluidics in Combination with a Compact Detection Device

Microfluidics offers several advantages, such as low cost, reduced reagent consumption, small sample volumes, increased reaction rates and faster analysis times for portable detection devices [51].

4.1.1. Fluorescence Resonance Energy Transfer-Based Systems

The detection of FRET biosensors in PON microfluidic devices has previously relied on using a fluorescence microscope to measure FRET signals. For example, Son et al. developed a microfluidic protease sensing system for monitoring cancer cell function through the release of the cancer-related protease Matrix metalloproteinase-9 (MMP9) [52]. Micro wells containing a hydrogel incorporating a FRET-linked peptide were integrated into a configurable Polydimethylsiloxane (PDMS) microfluidic device and visualized using a fluorescence microscope (Figure 2a). Cancer cells in suspension were trapped by immobilized antibodies in the hydrogel and the activity of MMP9 secreted from the trapped cells was monitored by following the cleavage of the FRET-linked peptides.

Cao et al. [54] developed a high-throughput multi-channel microfluidic chip to detect the interaction of fluorescently labeled aptamers with cancer cells. The surface of the chip was coated with graphene oxide (GO), which non-covalently binds to and quenches the fluorescence of the organic aptamers. Incubation of cancer cells on the chip led to the release of the fluorescent aptamers from the GO coating and recovery of the previously quenched fluorescence (Figure 2b). A single wavelength was used to measure the quantity of released aptamers.

A more compact and miniaturized sensing device for FRET signals used a laser to irradiate the microfluidic chip and optical fibers to deliver the emitted photons to photon-multiplier tubes (PMTs) [53]. This was used to detect three cancer markers simultaneously. Multiplexing was achieved by labeling three different aptamers with quantum dots having distinct emission spectra (Figure 2b). These were bound to GO-coated chips as in the single marker system. Signals were acquired by PMTs equipped with optical band-pass filters tailored to the different quantum-dot emission spectra. This system was capable of quantifying cancer markers in nanoliter-sized droplets of serum. The total assay time was 3 h. This might be too time-consuming to be considered as a classic point-of-need test; however, it could proof useful as a cancer diagnosis tool in remote settings if appropriate follow-up treatments are available. FRET quenching, as used here, does not have the advantages of being ratiometric, as only the activated fluorescence of the donor can be measured. On the other hand, the assay follows a "lights on" format, which is inherently more sensitive than "lights off".

4.1.2. Bioluminescence-Resonance-Energy-Transfer-Based Systems

Replacing FRET with BRET as the transduction modality markedly decreases optical noise and also means that a source of illumination, such as a laser, is no longer required. The BRET^2 system, comprising a variant of the *Renilla* luciferase, such as RLuc8, coupled to GFP^2, a large Stokes-shift variant of the Green Fluorescent Protein, is a highly sensitive RET tool. It exhibits an unusually large Förster distance [28,29] that can increase detection sensitivities, particularly when the radius of the biological recognition element is significantly greater than 1 nm. However, BRET^2 exhibits low and transient bioluminescence [58] that has to be accommodated for.

Our group initially demonstrated a proof of principle by using a BRET^2 thrombin sensor deployed in a PDMS microfluidic channel, with the BRET signal relayed through a microscope objective to two filter-equipped PMTs [59]. Subsequently, the microscope was dispensed with and ratiometric BRET^2 were made in micro liter volumes, using two filter-equipped PMTs combined with fiber optics, in close contact with the microfluidic device [60]. Although this device achieved the measurement of thrombin protease activities in buffer [60] and maltose detection in beer samples [61], it still lacked true point-of-need capability, as it was bulky and did not support on-chip incubation.

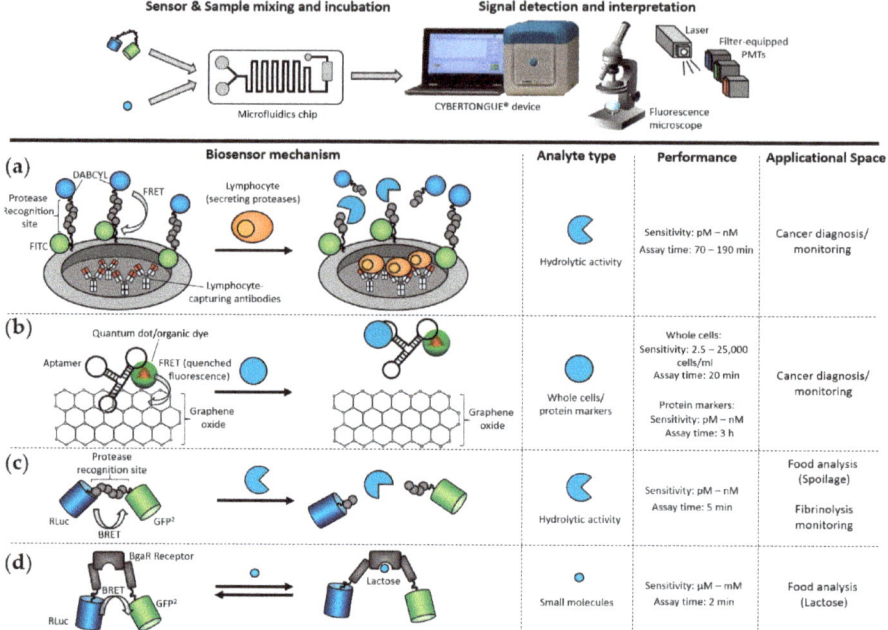

Figure 2. Examples of point-of-need (PON)-suitable applications using compact RET detection devices in combination with microfluidics. In one example, Bioluminescence Resonance Energy Transfer (BRET)-based biosensors are run on a microfluidics chip integrated into a compact device containing micro photon multiplier tubes (μPMTs). In other examples, FRET biosensor signals are recorded by using a fluorescence microscope or laser excitation followed by detection using PMTs. (**a**) Lymphocytes secreting Matrix metalloproteinase-9 (MMP9) are trapped by antibodies in a micro well located on a microfluidic chip. The peptides are labeled with Fluorescein isothiocyanate (FITC) and 4-(dimethylaminoazo)benzene-4-carboxylic acid (DABCYL), and they contain MMP9-specific cleavage sites. These are immobilized close to the micro wells, to detect any MMP9 activity released by the cells [52]. (**b**) Aptamers, labeled with quantum dots (QDs), specific for cancer-related cells or protein markers are attached on a graphene monoxide layer. Binding of the target cells or proteins to their specific aptamer results in a release of the aptamer from graphene oxide, activating the fluorescence signal of the quantum dot/organic dye [53,54]. (**c**) CYBERTONGUE® protease biosensors consist of the Renilla luciferase RLuc8 connected through a peptide linker, containing specific recognition sites for the target protease, to the Green Fluorescent Protein variant GFP². Proteolytic activity exerted on the connecting peptide results in the dissociation of GFP² from RLuc8, leading to a profound change in BRET ratio [48,55,56]. (**d**) The CYBERTONGUE® lactose biosensor consists of a lactose-binding protein tagged with RLuc8 and GFP² that undergoes a conformational change upon binding to lactose [57]. Binding of lactose results in the distancing of the two BRET components, thereby changing the BRET ratio.

In order to shrink the device's footprint, micro photon multiplier tubes, instead of large and energy-intensive valve-based PMTs, were placed directly above and below the sample detection chamber. This was implemented within a controlled microenvironment enabled by a thermoelectric block bringing the concept of a compact table top device, termed the Cybertongue® device, to fruition [48] (Figure 3a). The Cybertongue® device is a microfluidics-based platform that can run a variety of homogeneous sensing applications tailored to different analyte types with assay times of 1–10 min. The device combines many of the advantages of the aforementioned examples, such as small sample volumes, ratiometric RET signal measurements, a miniaturized device and rapid analysis times.

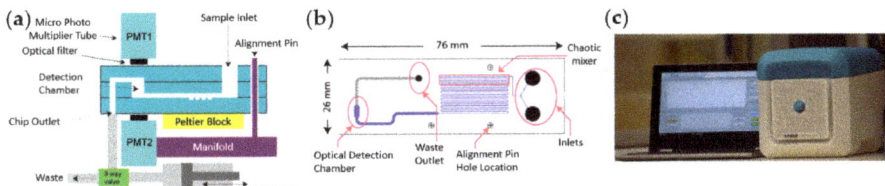

Figure 3. Overview of the Cybertongue® BRET analysis device: (**a**) functional schematic of the measurement device, (**b**) schematic design of a microfluidic chip used for protease assays and (**c**) image of compact microfluidics device with closed lid. Figure was taken from Weihs et al. [48], with permission.

The microfluidic channels are etched into a reusable glass chip (Figure 3b) that can be used repeatedly in an on-site setting, benefitting from the fact that glass is relatively inert to a variety of chemicals [62]. Chips with different microfluidic architectures can be inserted into the device, according to the type of assay. Options include a chaotic mixer, followed by serpentine channels, to improve fluid mixing on chip, or an incubation chamber for performing analyte–biosensor pre-incubations, if required.

Captured BRET signals are automatically interpreted via a Bluetooth-connected laptop, which further minimizes handling steps, from sample mixing to results. This setup has allowed the fabrication of a compact sensing system weighing 6.5 kg, suitable for on-site testing (Figure 3c).

4.1.3. Chemiluminescence-Resonance-Energy-Transfer-Based Systems

CRET relies on light emission generated by a redox reaction between luminol and hydrogen peroxide in the presence of a suitable catalyst, such as horse radish peroxidase or metal ions. Its advantages are similar to those of BRET, as it does not require an external illumination source, reducing optical noise, and detection devices can be easily miniaturized [63]. Homogeneous CRET-based assays have been described for the detection of several analytes, including ochratoxin A [64] and thrombin [65,66], using aptamer-mediated analyte binding or for the analysis of estradiol levels [67], C-reactive Protein [68] and Neuron-specific enolase [69] using antibody-based approaches.

However, while a range of commercially available on-site tests using chemiluminescence without RET do exist (allergy tests, flu tests, forensics, air pollutants, etc. [70]), the integration of CRET into suitable on-site testing systems has not been reported. One of the factors that may contribute to this lack of progress is that a number of biologically relevant molecules suppress CRET, such as human serum components [67]; biogenic amines; and thiols, amino acids, organic acids, and steroids [71]. The removal of these molecules via microchip electrophoresis prior analysis could resolve detection issues [71], but potentially adds complexity to a potential on-site test.

4.2. RET Detection Using Paper-Based Analytical Devices (PADs) in Combination with a Digital Camera

An alternative pathway to PON-compatible RET technologies is based on paper-based analytical devices (PADs). The low cost, portability and accessibility in low-resource settings of PADs has drawn considerable interest in recent years. Paper-based devices eliminate bulky instrumentation, such as plate readers combined with easy-to-carry-out procedures.

Filter or chromatography paper by Whatman are commonly used for paper-based assays. However, depending on the application, other paper selections can be of importance in the construction of paper-based sensing devices, as differences in porosity, hydrophobicity, pore size, thickness, fiber structure or grammage (mass per unit area) can affect performance of the supported biosensor system [72].

4.2.1. Bioluminescence-Resonance-Energy-Transfer-Based Systems

Antibodies are important biomarkers for the diagnosis and surveillance of fast-evolving infectious diseases and life-threatening allergies. Due to the urgency associated with such medical conditions, PON detection of antibody biomarkers could offer rapid answers and better health outcomes. An example of PON antibody detection was demonstrated for LUMABs (LUMinescent AntiBody Sensor) BRET-based biosensors (Figure 4a) [73,74]. These biosensors incorporate the highly luminescent luciferase NanoLuc [75] that enables BRET signal detection beyond a compact and controlled environment, such as in the Cybertongue® device. To directly quantify antibody in blood plasma, the authors were first able to carry out the homogenous BRET assay in solution, using a smartphone to measure the photon output within 20 min. To further improve the applicability of the test at PON, they then developed and applied the antibody assay on a paper support [74]. This fully integrated paper-based analytical device is composed of vertically assembled functionalized layers that facilitate blood plasma separation. This reduces the handling steps and simply requires the application of a drop of blood on the device. Similar to the corresponding homogenous assay, the results are obtained within 20 min, following photographic analysis with a smartphone. This system enables highly sensitive measurements of three antibodies against HIV, influenza and dengue, in blood, at low nanomolar concentrations (2.8 nM–19.3 nM detection limits). However, the LUMABS design is only applicable to antibodies that recognize linear epitopes, excluding the majority of antibodies that bind to conformational epitopes [76]. LUMABS were further developed to enable the detection of small molecules, such as dinitrophenol or creatinine, by the introduction of non-natural amino acids (pAzF) in place of the linear epitopes [77] (Figure 4b). This strategy can also be used to sense antibodies binding to conformational epitopes, if those are small molecules. However, paper-based detection of pAzF-LUMABS biosensors has not been reported yet.

A different sensor design also utilizing the NanoLuc luciferase, termed Luciferase-based indicators of drugs (LUCIDs), was followed by Johnsson and co-workers for the monitoring of small molecule drugs [79] and later applied for use on paper-based strips, in combination with a digital camera [80,81]. For this purpose, lyophilized biosensors were mounted onto filter paper and liquid samples, such as human serum or whole blood including the luciferase substrate, were added to the paper strip. The reaction between the biosensor and the analyte occurs on paper, and the bright luminescence of NanoLuc is detected by taking a photograph of the paper strip, followed by a software-guided photon analysis. The combination of the lyophilized stabilized biosensor system and the use of low-tech, easily accessible analytical equipment, such as a digital camera, facilitates the translation of such a technology to point-of-need. Following this first LUCIDs PON application, a range of LUCID biosensors were developed for the detection of analytes, such as phenylalanine or glutamate [80] and the clinical drugs methotrexate, theophylline and quinine (Figure 4c) [81]. The detection of such analytes in whole blood at PON is of high interest due to their clinical relevance for monitoring children and pregnant women suffering from phenylketonuria.

MicroRNAs (miRNAs) provide vital information about many diseases and are of great interest in the diagnosis and monitoring of cancer [90]. Analyses of miRNAs usually require a time-consuming amplification step and a PCR instrument. Wu and co-workers recently developed a BRET-based point-of-care suitable technique for the detection of miRNAs [82]. This was achieved by coupling a paper-based isothermal rolling circle DNA amplification with detection by biosensors incorporating NanoLuc and mNeonGreen fused with DNA sequence-specific Zinc Finger Proteins (ZFPs). Human serum can be applied on a paper disc containing lyophilized components required to amplify the target miRNAs through circular single-stranded DNA amplicons. Only the presence of the target miRNAs, acting as primers, enables the amplification of the DNA amplicons. Subsequently, another paper disc, one that contains miRNA complementary oligonucleotides and BRET biosensors, is applied on top of the amplification disc. If miRNAs were present in the sample, the complementary oligonucleotides form double-stranded DNAs with the amplicons that are, in turn, recognized by NanoLuc-ZFP and mNeonGreen-ZFP (Figure 4d). In absence of the miRNAs, both fusions remain dispersed and no BRET can be observed. Signals were detected via a smartphone camera. The speed of point-of-care diagnosis can be less relevant to cancer diagnoses, since these usually do not require quick diagnoses. However, such a system might be of value when it important to monitor biomarkers during treatment visits or to diagnose in remote and resource-limited settings.

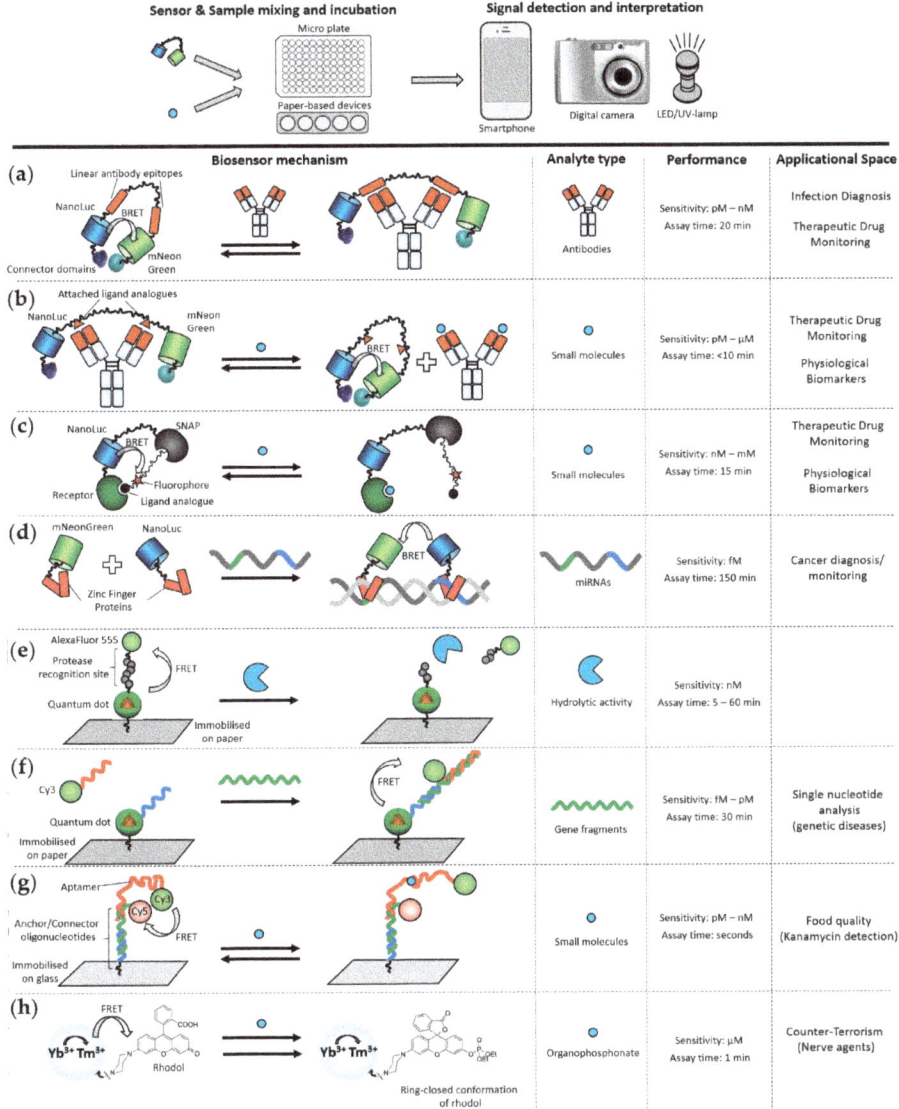

Figure 4. Examples of PON-suitable applications using a digital camera or smartphone in combination with micro plates or paper-based devices. BRET-based biosensors are spotted on paper-based analytical devices (PADs), and signals are recorded with a digital camera or smart phone. FRET biosensors require an additional source of excitation, such as a light-emitting diode (LED) or UV-lamp. (**a**) LUMABS biosensors (LUMinescent AntiBody Sensor) biosensors (LUMinescent AntiBody Sensor) are comprised

of the luciferase NanoLuc and the fluorescent protein mNeonGreen connected through a linker containing linear epitopes for antibodies of interest. In the absence of the antibody of interest, NanoLuc and mNeonGreen dimerize through the connector domains [73,74,78]. (**b**) LUMABs were modified by replacing linear epitopes with unnatural amino acids acting as a chemical handle to introduce analogues of analytes of interest. An antibody binding to these analogues is introduced, separating NanoLuc and mNeonGreen. In the presence of the analyte, antibodies preferentially bind to the analyte instead of its analogues incorporated in the LUMAB biosensor [77]. (**c**) LUCIDs (luciferase-based indicators of drugs) are protein fusions comprising NanoLuc, a receptor protein for the drug of interest and the self-labeling enzyme SNAP. A SNAP-functionalized organic dye Cy3 is attached to an analyte analogue, which is covalently incorporated by the SNAP protein. In the presence of the analyte, the receptor preferentially binds the analyte over its SNAP–Cy3-bound analogue [79–81]. (**d**) If target miRNAs are present in a sample, miRNA templates are amplified through a rolling circle amplification (not illustrated). Complementary oligonucleotides form double-stranded DNAs that are recognized by fusions of NanoLuc and mNeonGreen with zinc finger proteins that specifically bind to different but nearby sequences [82]. (**e**) Quantum dot (QD)–organic fluorescent dye conjugates joined by a peptide-containing protease-specific recognition site are immobilized on paper. In the absence of proteolytic activity, FRET occurs between the QD and the organic dye, resulting in a yellow/orange emission signal. If the peptide is cleaved due to the proteolytic activity of the protease of interest, the dye diffuses away from the QD, leading to a green emission from the QDs [83]. (**f**) Paper-immobilized quantum dot–oligonucleotides and free Cy3–oligonucleotides contain different DNA segments complementary to the target gene fragment. In a sandwich format, the target gene serves as a hybridization bridge for the QD–oligonucleotide and Cy3–oligonucleotide, which in turn enables FRET between QD and Cy3 [84–87]; (**g**) A Cy3-labeled kanamycin-specific aptamer partially hybridizes to an anchor/connector oligonucleotide immobilized on glass. The connector oligonucleotide is conjugated to Cy5. Binding of kanamycin spatially separates Cy3 from Cy5 components, leading to a lower FRET efficiency [88]. (**h**) An upconversion nanoparticle (UCNP) consisting of ytterbium (Yb^{3+}) and thulium (Tm^{3+}) is conjugated to the organic dye rhodol. FRET occurs between the UCNP and rhodol, while organophosphonates perform a nucleophilic attack on rhodol, inactivating it as a FRET acceptor [89].

4.2.2. Fluorescence-Resonance-Energy-Transfer-Based Systems

A range of FRET biosensors have been developed for paper-based devices and analyzed by using a digital camera. For instance, Petrayayeva et al. developed a paper-based assay to monitor protease activities at the PON or in low-resources settings [83]. The biosensor consisted of on-paper spots of quantum dots (QDs) and organic fluorescent dyes covalently linked by a peptide sequence specific to the protease of interest. In the absence of proteolytic activity (Figure 4e), FRET occurs between the QD and the organic dye, resulting in a yellow/orange emission signal. If the peptide is cleaved due to the proteolytic activity of the protease of interest, the dye diffuses away from the QD, leading to a green emission from the QDs. To bring this sensing method suitable to the PON, the authors demonstrated that a battery-powered LED as the excitation source and a smart phone can be used to quantitatively determine protease activities within 5 min.

Krull and co-workers developed different sensing systems deployed at PON, using a paper-based design. The biosensors are based on the transduction of nucleic acid hybridization [84,85,87,91,92] and both a direct and a sandwich format assay were developed, both based on paper-immobilized quantum dot-oligonucleotide conjugates (Figure 4f, sandwich assay format shown) [92,93]. Hybridization events in both formats brings the QD into close proximity with Cy3 [92]. Although the technology was first developed using lab-based epifluorescence, the researchers then moved to more PON-friendly equipment. They reported the adaptation of the test to paper substrate, combined with a handheld UV lamp as the excitation source and an iPad camera for the ratiometric fluorescence emission analyses. In comparison with the use of sophisticated laboratory equipment—in this case, an epifluorescence microscope—the authors reported a sensitivity approximately one order of magnitude less sensitive when using the portable equipment. The PON test described above was also used as the basis of rapid diagnostic tests, to detect a single nucleotide polymorphism indicative of spinal muscular atrophy [87] and a three-base replacement used in the diagnosis of cystic fibrosis [86].

Krull and co-workers also used a similar oligonucleotide-based system where a quantum dot was conjugated to a single-stranded DNA aptamer that specifically binds the cancer biomarker protein epithelial cell adhesion molecule (EpCAM), immobilized on paper [84]. A conjugate of Cy3 with an oligonucleotide complementary to the aptamer is added to the immobilized construct. FRET between the QD and an added Cy3-DNA conjugate could be observed in the absence of EpCAM. When EpCAM is present, it displaces the Cy3-DNA conjugate from the aptamer, resulting in a loss of FRET signal (Figure 4g). Multiple biosensors can be delivered in this way at the PON, including smaller aptasensors, which have been used to detect antibiotics in milk [88]; however, these approaches have not yet been adopted commercially.

4.2.3. FRET Incorporating Upconversion Nanoparticles (UCNPs)

Luminescence resonance energy transfer (LRET) is a type of RET between upconverting nanoparticles (UCNPs), such as lanthanides, and a FRET acceptor. Translation of LRET biosensors for PON applications was also achieved by using PADs by Krull and co-workers [94]. The authors used UCNPs as the FRET donor and QDs as the acceptors to build a multiplexed paper-based assay for the simultaneous detection of three diagnostic markers for the bacterium *E.coli* [95]. Although multiplexing of the biosensor devices was achieved on paper, analyses of the paper strips were carried out by using an epi-fluorescence microscope, which is incompatible with PON applications. This highlights the limitations sometimes encountered with FRET-dependent systems that have only been demonstrated by using bulky, sophisticated laboratory equipment.

In another example of UCNP-based sensing, Wang and co-workers successfully adapted a nanosensor for PON detection of an organophosphonate nerve agent (Figure 4h) [89]. The biosensor development was initially carried out by using sophisticated laboratory equipment, after which the authors also reported that the exposure of the paper-strip biosensor to the nerve agent of interest leads to the emission of blue light that is visible with the naked eye.

4.2.4. FRET-Based Systems Using Time Resolved Measurements

Time-resolved FRET (TR-FRET) involves measuring the fluorescence lifetimes of the FRET components instead of taking ratiometric measurements between donor and acceptor signal. Traditionally, TR-FRET has required heavy and complex lab-based machines, such as plate readers. For instance, commercially available assays, such as BRAHM's TRACE® assays, are applied on the KRYPTOR device [96], which is a 54 kg benchtop device, which is less suitable for on-site testing. Recently however, ProciseDx has produced a more PON-friendly device [97]. The shoe-box-sized device will allow the analysis of complex fluid samples, such as finger prick whole blood that can be inserted into the device via cartridges that contain the biosensor reagents, followed by their automatic analysis. Another user-friendly feature lies within the automatic mixing of sample and biosensor in the cartridge, removing a source of error through pipetting. No assays for the ProciseDx are commercially available at this point, but it has been announced that the first assay will be a 5 min test to quantitate the inflammatory marker C-reactive protein [98].

The translation of TR-FRET from bulky bench-top devices into reliable portable detection devices would open up a myriad of options for on-site applications, as a wide range of TR-FRET assays is already commercially available [96,99] A range of other TR-FRET assays have also been described, including for the diagnosis of sepsis, cancer, disorders, virus infections [100] or celiac disease [101] (Figure 5).

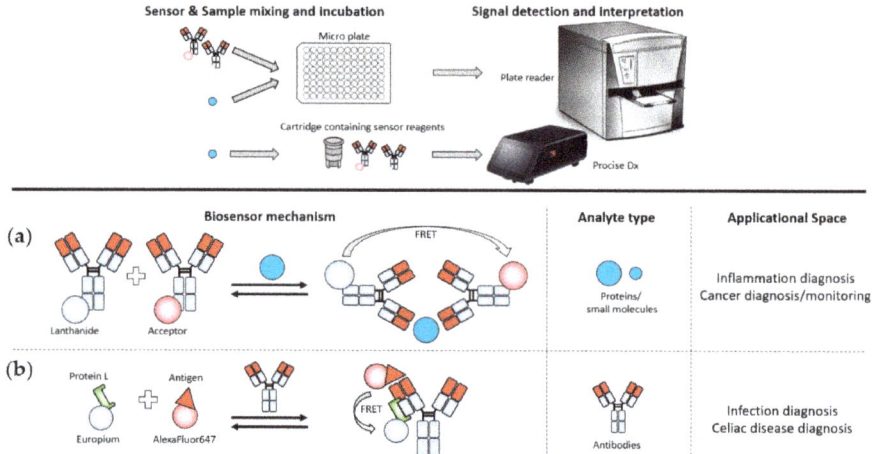

Figure 5. Examples of Lanthanide-FRET (LRET) applications that are potentially suitable for on-site testing. (**a**) Commonly applied TR-FRET technologies rely on a sandwich-based homogeneous assay, where two antibodies targeting different epitopes of the analyte are labeled with a lanthanide energy donor, while the other is labeled, with an organic dye or fluorescent protein, as the energy acceptor. Binding of both antibodies enables FRET between the lanthanide and acceptor which is measured by their altered fluorescence lifetimes. (**b**) Protein L, an antibody light-chain-binding protein [102], is labeled with Europium. Antigens to an antibody of interest are labeled with the organic dye AlexaFluor647 (LFRET) [103]. If a sample contains antibodies against the antigen–dye fusion, FRET occurs between Protein-L-Europium bound to the light chain of the antibody and the Antigen–AlexaFluor647 fusion. Image of the ProciseDx device was used with permission from ProciseDx.

5. Conclusions and Perspectives for RET-Based PON Systems

In recent years, we have seen tremendous progress in translating a range of laboratory-based RET biosensors into detection methods suitable for point-of-need testing. Improvements have occurred in multiple areas. These involve improved optical donors and acceptors, expanding the range of molecular architectures so as to creatively link an ever-expanding variety of biological recognition elements, developing improved assay matrices and miniaturizing suitable detection devices. While it is unlikely to see handheld all-in-one detection tools, such as the glucose meter, any time soon, due to their necessity of at least a photon-collecting component, RET-assays are quantitative and can be highly sensitive. This opens up their application to areas where trace-level analyses, such as for protease activity or antibiotics, and/or where quantitative measurements are needed rather than simply flagging the presence of an analyte at some (difficult to standardize) threshold. In the clinical area, such requirements often occur in the case of measuring inflammatory or metabolic markers.

A majority of RET-based techniques that can be applied at the point-of-need are amplification-free, allowing tests that deliver results within a few minutes. Gold standards such as PCRs for the detection of miRNAs or gene fragments, HPLC for the analysis of small molecules and ELISAs for the quantitation of proteins and small molecules are still too time-consuming, are user unfriendly and require specialized instruments to be deployed at the point-of-need, despite ongoing attempts to mitigate these disadvantages. On the other hand, RET-techniques hold the potential to deliver a step-change improvement in the speed and simplicity of point-of-need testing. Ultimately, the use-case is what determines the assay-time requirements. For example, clearing a food-processing line of allergens before commencing with a different product demands rapid, sensitive and quantitative results. Similarly, certain medical conditions, such as drug intoxication or infection, may benefit from test results that can be delivered within minutes.

Although no RET-based point-of-need applications are fully commercial yet, we see a clear drive towards developing FRET methods to meet the requirements at the PON. However, the inherent requirement of FRET for bulky illumination sources, such as fluorescence microscopes [52,54], lasers [53,104], arc lamps [85] or LEDs [83], is hindering the development of miniaturized or portable devices. In addition, FRET-based techniques suffer from high background levels in most complex samples, especially those containing autofluorescent molecules. The use of infrared-shifted FRET systems, enabling excitation wavelengths that do not trigger autofluorescent molecules, or applying TR-FRET can circumvent issues stemming from high background levels. Especially, TR-FRET is a rapid and sensitive alternative to sensitized FRET techniques, which could be a game changer at the point-of-need when smaller detection devices, such as the ProciseDx, with an easy-to-handle cartridge system, become widely available.

Striking progress has been seen in the field of RET biosensors that do not require an external light source, especially for biosensors incorporating BRET as the transduction modality. The Cybertongue® technology is approaching PON acceptance through a solid state BRET detection system within a microfluidics- and temperature-controlled reaction environment [48]. This system can be applied at the point-of-need, for instance in food production facilities, with minimum sample preparation and limited need for operator training while maintaining highly reproducible results. Another approach focuses on truly portable sensing systems that are applied on PADs and analyzed by using a smart phone [73,74,81]. Such tools lend themselves to applications in resource-limited or highly time critical settings, such as for certain medical conditions; however, it remains to be seen if they can meet regulatory standards for sensitivity and reliability in clinical use. We expect their developers to focus strongly on demonstrating these features in the short-to-medium term.

With all of these development pathways occurring in parallel, we hope to see further rapid advances in the field, particularly as the advantages of different systems are combined and adapted. Moreover, we are reasonably optimistic that the next five years will see some of the technologies we have briefly reviewed herein taking their place in the clinic and the factory, alongside well-established LF and EC platforms.

Author Contributions: Conceptualization, F.W., A.A., S.T. and K.C.; writing—original draft preparation, F.W., A.A. and K.C.; writing—review and editing, F.W., A.A., S.T. and K.C. All authors have read and agreed to the published version of the manuscript.

Funding: F.W. is funded by the CSIRO Probing Biosystems Future Science Platform. A.A. and K.C. are funded by the CSIRO Health & Biosecurity Business Unit.

Institutional Review Board Statement: Not applicable.

Informed Consent Statement: Not applicable.

Data Availability Statement: Not applicable.

Acknowledgments: We would like to thank Karolina Petkovic and Wayne Leifert for their helpful suggestions in the preparation of the manuscript.

Conflicts of Interest: K.C. and S.T. are listed as inventors in patents submitted by the CSIRO related to the Cybertongue® technology. S.T. is a founder of PPB Technology, which has licensed certain commercial rights to those patents.

References

1. McPartlin, D.A.; O'Kennedy, R.J. Point-of-care diagnostics, a major opportunity for change in traditional diagnostic approaches: Potential and limitations. *Expert Rev. Mol. Diagn.* **2014**, *14*, 979–998. [CrossRef]
2. Shanmugakani, R.K.; Srinivasan, B.; Glesby, M.J.; Westblade, L.F.; Cárdenas, W.B.; Raj, T.; Erickson, D.; Mehta, S. Current state of the art in rapid diagnostics for antimicrobial resistance. *Lab Chip* **2020**, *20*, 2607–2625. [CrossRef]
3. Busa, L.S.A.; Mohammadi, S.; Tokeshi, M.; Ishida, A.; Tani, H.; Tokeshi, M. Advances in Microfluidic Paper-Based Analytical Devices for Food and Water Analysis. *Micromachines* **2016**, *7*, 86. [CrossRef]

4. Choi, J.R.; Yong, K.W.; Choi, J.Y.; Cowie, A.C. Emerging Point-of-care Technologies for Food Safety Analysis. *Sensors* **2019**, *19*, 817. [CrossRef]
5. Pontius, K.; Semenova, D.; Silina, Y.E.; Gernaey, K.V.; Junicke, H. Automated Electrochemical Glucose Biosensor Platform as an Efficient Tool Toward On-Line Fermentation Monitoring: Novel Application Approaches and Insights. *Front. Bioeng. Biotechnol.* **2020**, *8*, 15. [CrossRef] [PubMed]
6. De Paepe, B.; Peters, G.; Coussement, P.; Maertens, J.; De Mey, M. Tailor-made transcriptional biosensors for optimizing microbial cell factories. *J. Ind. Microbiol. Biotechnol.* **2017**, *44*, 623–645. [CrossRef] [PubMed]
7. Wang, S.; Hossack, J.A.; Klibanov, A.L. From Anatomy to Functional and Molecular Biomarker Imaging and Therapy: Ultrasound Is Safe, Ultrafast, Portable, and Inexpensive. *Investig. Radiol.* **2020**, *55*, 559–572. [CrossRef] [PubMed]
8. Wald, L.L.; Ms, P.C.M.; Witzel, T.; Stockmann, J.P.; Cooley, C.Z. Low-cost and portable MRI. *J. Magn. Reson. Imaging* **2019**, *52*, 686–696. [CrossRef] [PubMed]
9. Crocombe, R.A. Portable Spectroscopy. *Appl. Spectrosc.* **2018**, *72*, 1701–1751. [CrossRef]
10. Broughton, J.P.; Deng, X.; Yu, G.; Fasching, C.L.; Servellita, V.; Singh, J.; Miao, X.; Streithorst, J.A.; Granados, A.; Sotomayor-Gonzalez, A.; et al. CRISPR–Cas12-based detection of SARS-CoV-2. *Nat. Biotechnol.* **2020**, *38*, 870–874. [CrossRef]
11. Wang, Y.; Zhou, J.; Li, J. Construction of Plasmonic Nano-Biosensor-Based Devices for Point-of-Care Testing. *Small Methods* **2017**, *1*, 1. [CrossRef]
12. Marks, H.; Schechinger, M.; Garza, J.; Locke, A.K.; Coté, G. Surface enhanced Raman spectroscopy (SERS) for in vitro diagnostic testing at the point of care. *Nanophotonics* **2017**, *6*, 681–701. [CrossRef]
13. Andryukov, B.G. Six decades of lateral flow immunoassay: From determining metabolic markers to diagnosing COVID-19. *AIMS Microbiol.* **2020**, *6*, 280–304. [CrossRef] [PubMed]
14. Busin, V.; Wells, B.; Kersaudy-Kerhoas, M.; Shu, W.; Burgess, S.T.G. Opportunities and challenges for the application of microfluidic technologies in point-of-care veterinary diagnostics. *Mol. Cell. Probes* **2016**, *30*, 331–341. [CrossRef]
15. Santos, V.O.; Pelegrini, P.B.; Mulinari, F.; Lacerda, A.F.; Moura, R.S.; Cardoso, L.P.V.; Bührer-Sékula, S.; Miller, R.N.G.; Molinari, H.B.C. Development and validation of a novel lateral flow immunoassay device for detection of aflatoxins in soy-based foods. *Anal. Methods* **2017**, *9*, 2715–2722. [CrossRef]
16. Vickers, S.; Bernier, M.; Zambrzycki, S.; Fernandez, F.M.; Newton, P.N.; Caillet, C. Field detection devices for screening the quality of medicines: A systematic review. *BMJ Glob. Heal.* **2018**, *3*, e000725. [CrossRef]
17. Posthuma-Trumpie, G.A.; Korf, J.; Van Amerongen, A. Lateral flow (immuno)assay: Its strengths, weaknesses, opportunities and threats. A literature survey. *Anal. Bioanal. Chem.* **2009**, *393*, 569–582. [CrossRef]
18. Urusov, A.E.; Zherdev, A.V.; Dzantiev, B.B. Towards Lateral Flow Quantitative Assays: Detection Approaches. *Biosensors* **2019**, *9*, 89. [CrossRef]
19. De Puig, H.; Bosch, I.; Gehrke, L.; Hamad, K. Challenges of the Nano-Bio Interface in Lateral Flow and Dipstick Immunoassays. *Trends Biotechnol.* **2017**, *35*, 1169–1180. [CrossRef]
20. Bishop, J.D.; Hsieh, H.V.; Gasperino, D.J.; Weigl, B.H. Sensitivity enhancement in lateral flow assays: A systems perspective. *Lab Chip* **2019**, *19*, 2486–2499. [CrossRef]
21. Hsieh, H.V.; Dantzler, J.L.; Weigl, B.H. Analytical Tools to Improve Optimization Procedures for Lateral Flow Assays. *Diagnostics* **2017**, *7*, 29. [CrossRef] [PubMed]
22. Vashist, S.K.; Zheng, D.; Al-Rubeaan, K.; Luong, J.H.; Sheu, F.-S. Technology behind commercial devices for blood glucose monitoring in diabetes management: A review. *Anal. Chim. Acta* **2011**, *703*, 124–136. [CrossRef] [PubMed]
23. Anik, U. Electrochemical medical biosensors for POC applications. In *Medical Biosensors for Point of Care*; Narayan, R.J., Ed.; Woodhead Publ Ltd.: Cambridge, UK, 2017; Volume 118, pp. 275–292.
24. Vigneshvar, S.; Sudhakumari, C.C.; Senthilkumaran, B.; Prakash, H. Recent Advances in Biosensor Technology for Potential Applications—An Overview. *Front. Bioeng. Biotechnol.* **2016**, *4*, 11. [CrossRef] [PubMed]
25. Available online: www.sherlock.bio/platforms/sherlock/ (accessed on 5 January 2020).
26. Förster, T. Zwischenmolekulare Energiewanderung und Fluoreszenz. *Ann. Phys.* **1948**, *437*, 55–75. [CrossRef]
27. Bajar, B.T.; Wang, E.S.; Zhang, S.; Lin, M.Z.; Chu, J. A Guide to Fluorescent Protein FRET Pairs. *Sensors* **2016**, *16*, 1488. [CrossRef]

28. Dacres, H.; Michie, M.; Wang, J.; Pfleger, K.D.; Trowell, S.C. Effect of enhanced Renilla luciferase and fluorescent protein variants on the Förster distance of Bioluminescence resonance energy transfer (BRET). *Biochem. Biophys. Res. Commun.* **2012**, *425*, 625–629. [CrossRef]
29. Dacres, H.; Wang, J.; Dumancic, M.M.; Trowell, S.C. Experimental Determination of the Förster Distance for Two Commonly Used Bioluminescent Resonance Energy Transfer Pairs. *Anal. Chem.* **2010**, *82*, 432–435. [CrossRef]
30. Zhang, L.; Takahashi, Y.; Hsu, P.; Kollist, H.; Merilo, E.; Krysan, P.J.; Schroeder, J.I. FRET kinase sensor development reveals SnRK2/OST1 activation by ABA but not by MeJA and high CO2 during stomatal closure. *eLife* **2020**, *9*, 9. [CrossRef]
31. Weihs, F.; Peh, A.; Dacres, H. A red-shifted Bioluminescence Resonance Energy Transfer (BRET) biosensing system for rapid measurement of plasmin activity in human plasma. *Anal. Chim. Acta* **2020**, *1102*, 99–108. [CrossRef]
32. Sun, S.; Liu, Y.; Xia, J.; Wang, M.; Tang, R.; Lei, C.; Huang, Y.; Li, Y.; Yao, S. A semisynthetic fluorescent protein assembly-based FRET probe for real-time profiling of cell membrane protease functions in situ. *Chem. Commun.* **2019**, *55*, 2218–2221. [CrossRef]
33. Emmott, E.; Sweeney, E.P.; Goodfellow, I.G. A Cell-based Fluorescence Resonance Energy Transfer (FRET) Sensor Reveals Inter- and Intragenogroup Variations in Norovirus Protease Activity and Polyprotein Cleavage. *J. Biol. Chem.* **2015**, *290*, 27841–27853. [CrossRef] [PubMed]
34. Stoddart, L.A.; Vernall, A.J.; Bouzo-Lorenzo, M.; Bosma, R.; Kooistra, A.J.; De Graaf, C.; Vischer, H.F.; Leurs, R.; Briddon, S.J.; Kellam, B.; et al. Development of novel fluorescent histamine H1-receptor antagonists to study ligand-binding kinetics in living cells. *Sci. Rep.* **2018**, *8*, 1–19. [CrossRef] [PubMed]
35. Schihada, H.; Vandenabeele, S.; Zabel, U.; Frank, M.; Lohse, M.J.; Maiellaro, I. A universal bioluminescence resonance energy transfer sensor design enables high-sensitivity screening of GPCR activation dynamics. *Commun. Biol.* **2018**, *1*, 1–8. [CrossRef] [PubMed]
36. Golynskiy, M.V.; Rurup, W.F.; Merkx, M.A.W. Antibody Detection by Using a FRET-Based Protein Conformational Switch. *ChemBioChem* **2010**, *11*, 2264–2267. [CrossRef] [PubMed]
37. Ingram, J.M.; Zhang, C.; Xu, J.; Schiff, S.J. FRET excited ratiometric oxygen sensing in living tissue. *J. Neurosci. Methods* **2013**, *214*, 45–51. [CrossRef] [PubMed]
38. Yuan, L.; Lin, W.; Zheng, K.; Zhu, S. FRET-Based Small-Molecule Fluorescent Probes: Rational Design and Bioimaging Applications. *Accounts Chem. Res.* **2013**, *46*, 1462–1473. [CrossRef] [PubMed]
39. Weihs, F.; Wacnik, K.; Turner, R.D.; Culley, S.; Henriques, R.; Foster, S.J. Heterogeneous localisation of membrane proteins in Staphylococcus aureus. *Sci. Rep.* **2018**, *8*, 1–11. [CrossRef]
40. Margineanu, A.; Chan, J.J.; Kelly, D.J.; Warren, S.C.; Flatters, D.; Kumar, S.; Katan, M.; Dunsby, C.W.; French, P.M.W. Screening for protein-protein interactions using Förster resonance energy transfer (FRET) and fluorescence lifetime imaging microscopy (FLIM). *Sci. Rep.* **2016**, *6*, 28186. [CrossRef]
41. Park, S.-H.; Ko, W.; Lee, H.S.; Shin, I. Analysis of Protein–Protein Interaction in a Single Live Cell by Using a FRET System Based on Genetic Code Expansion Technology. *J. Am. Chem. Soc.* **2019**, *141*, 4273–4281. [CrossRef]
42. Clark, L.C., Jr.; Lyons, C. Electrode systems for continuous monitoring in cardiovascular surgery. *Ann. N. Y. Acad. Sci.* **1962**, *102*, 29–45. [CrossRef]
43. Weihs, F.; Dacres, H. Red-shifted bioluminescence Resonance Energy Transfer: Improved tools and materials for analytical in vivo approaches. *TrAC Trends Anal. Chem.* **2019**, *116*, 61–73. [CrossRef]
44. England, C.G.; Ehlerding, E.B.; Cai, W. NanoLuc: A Small Luciferase Is Brightening Up the Field of Bioluminescence. *Bioconjugate Chem.* **2016**, *27*, 1175–1187. [CrossRef] [PubMed]
45. Kaskova, Z.M.; Tsarkova, A.S.; Yampolsky, I.V. 1001 lights: Luciferins, luciferases, their mechanisms of action and applications in chemical analysis, biology and medicine. *Chem. Soc. Rev.* **2016**, *45*, 6048–6077. [CrossRef] [PubMed]
46. Rodriguez, E.A.; Campbell, R.E.; Lin, J.Y.; Lin, M.Z.; Miyawaki, A.; Palmer, A.E.; Shu, X.; Zhang, J.; Tsien, R.Y. The Growing and Glowing Toolbox of Fluorescent and Photoactive Proteins. *Trends Biochem. Sci.* **2017**, *42*, 111–129. [CrossRef] [PubMed]
47. Machleidt, T.; Woodroofe, C.C.; Schwinn, M.K.; Méndez, J.; Robers, M.B.; Zimmerman, K.; Otto, P.; Daniels, D.L.; Kirkland, T.A.; Wood, K.V. NanoBRET—A Novel BRET Platform for the Analysis of Protein–Protein Interactions. *ACS Chem. Biol.* **2015**, *10*, 1797–1804. [CrossRef]

48. Weihs, F.; Gel, M.; Wang, J.; Anderson, A.; Trowell, S.; Dacres, H. Development and characterisation of a compact device for rapid real-time-on-chip detection of thrombin activity in human serum using bioluminescence resonance energy transfer (BRET). *Biosens. Bioelectron.* **2020**, *158*, 112162. [CrossRef]
49. Lakowicz, J.R.; Masters, B.R. *Principles of Fluorescence Spectroscopy*, 3rd ed.; Kluwer-Plenum: New York, NY, USA, 2008; p. 029901. [CrossRef]
50. Wang, D.; Zhang, J.; Lin, Q.; Fub, L.; Zhang, H.; Yang, B. Lanthanide complex/polymer composite optical resin with intense narrow band emission, high transparency and good mechanical performance. *J. Mater. Chem.* **2003**, *13*, 2279–2284. [CrossRef]
51. Whitesides, G.M. The origins and the future of microfluidics. *Nat. Cell Biol.* **2006**, *442*, 368–373. [CrossRef]
52. Son, K.J.; Shin, D.-S.; Kwa, T.; Gao, Y.; Revzin, A. Micropatterned Sensing Hydrogels Integrated with Reconfigurable Microfluidics for Detecting Protease Release from Cells. *Anal. Chem.* **2013**, *85*, 11893–11901. [CrossRef]
53. Zhang, Y.; Ye, W.; Yang, C.; Xu, Z.-R. Simultaneous quantitative detection of multiple tumor markers in microfluidic nanoliter-volume droplets. *Talanta* **2019**, *205*, 120096. [CrossRef]
54. Cao, L.; Cheng, L.; Zhang, Z.; Wang, Y.; Zhang, X.; Chen, H.; Liu, B.; Zhang, S.; Kong, J. Visual and high-throughput detection of cancer cells using a graphene oxide-based FRET aptasensing microfluidic chip. *Lab Chip* **2012**, *12*, 4864–4869. [CrossRef] [PubMed]
55. Dacres, H.; Wang, J.; Anderson, A.; Trowell, S.C. A rapid and sensitive biosensor for measuring plasmin activity in milk. *Sens. Actuators B Chem.* **2019**, *301*, 127141. [CrossRef]
56. Dacres, H.; Trowell, S.C. Protease Sensor Molecules. Australian Patent WO/2018/085895, 17 May 2018.
57. Caron, K.; Trowell, S.C. Highly Sensitive and Selective Biosensor for a Disaccharide Based on an AraC-Like Transcriptional Regulator Transduced with Bioluminescence Resonance Energy Transfer. *Anal. Chem.* **2018**, *90*, 12986–12993. [CrossRef] [PubMed]
58. Pfleger, K.D.G.; Dromey, J.R.; Dalrymple, M.B.; Lim, E.M.; Thomas, W.G.; Eidne, K.A. Extended bioluminescence resonance energy transfer (eBRET) for monitoring prolonged protein–protein interactions in live cells. *Cell. Signal.* **2006**, *18*, 1664–1670. [CrossRef]
59. Wu, N.; Dacres, H.; Anderson, A.; Trowell, S.C.; Zhu, Y. Comparison of Static and Microfluidic Protease Assays Using Modified Bioluminescence Resonance Energy Transfer Chemistry. *PLoS ONE* **2014**, *9*, e88399. [CrossRef] [PubMed]
60. Le, N.; Gel, M.; Zhu, Y.; Wang, J.; Dacres, H.; Anderson, A.R.; Trowell, S.C. Sub-nanomolar detection of thrombin activity on a microfluidic chip. *Biomicrofluidics* **2014**, *8*, 64110. [CrossRef]
61. Le, N.; Gel, M.; Zhu, Y.; Dacres, H.; Anderson, A.; Trowell, S. Real-time, continuous detection of maltose using bioluminescence resonance energy transfer (BRET) on a microfluidic system. *Biosens. Bioelectron.* **2014**, *62*, 177–181. [CrossRef]
62. Lee, J.N.; Park, A.C.; Whitesides, G.M. Solvent Compatibility of Poly(dimethylsiloxane)-Based Microfluidic Devices. *Anal. Chem.* **2003**, *75*, 6544–6554. [CrossRef]
63. Alahmad, W.; Uraisin, K.; Nacapricha, D.; Kaneta, T. A miniaturized chemiluminescence detection system for a microfluidic paper-based analytical device and its application to the determination of chromium(iii). *Anal. Methods* **2016**, *8*, 5414–5420. [CrossRef]
64. Jo, E.-J.; Mun, H.; Kim, S.-J.; Shim, W.B.; Kim, M.-G. Detection of ochratoxin A (OTA) in coffee using chemiluminescence resonance energy transfer (CRET) aptasensor. *Food Chem.* **2016**, *194*, 1102–1107. [CrossRef]
65. Mun, H.; Jo, E.-J.; Li, T.; Joung, H.-A.; Hong, D.-G.; Shim, W.B.; Jung, C.; Kim, M.-G. Homogeneous assay of target molecules based on chemiluminescence resonance energy transfer (CRET) using DNAzyme-linked aptamers. *Biosens. Bioelectron.* **2014**, *58*, 308–313. [CrossRef] [PubMed]
66. Qin, G.; Zhao, S.; Huang, Y.; Jiang, J.; Liu, Y.-M. A sensitive gold nanoparticles sensing platform based on resonance energy transfer for chemiluminescence light on detection of biomolecules. *Biosens. Bioelectron.* **2013**, *46*, 119–123. [CrossRef] [PubMed]
67. Zhao, S.; Liu, J.; Huang, Y.; Liu, Y.-M. Introducing chemiluminescence resonance energy transfer into immunoassay in a microfluidic format for an improved assay sensitivity. *Chem. Commun.* **2012**, *48*, 699–701. [CrossRef] [PubMed]
68. Lee, J.S.; Joung, H.-A.; Kim, M.-G.; Park, C.B. Graphene-Based Chemiluminescence Resonance Energy Transfer for Homogeneous Immunoassay. *ACS Nano* **2012**, *6*, 2978–2983. [CrossRef]

69. Yang, T.; Vdovenko, M.; Jin, X.; Sakharov, I.Y.; Zhao, S. Highly sensitive microfluidic competitive enzyme immunoassay based on chemiluminescence resonance energy transfer for the detection of neuron-specific enolase. *Electrophoresis* **2014**, *35*, 2022–2028. [CrossRef] [PubMed]
70. Park, J.Y.; Kricka, L.J. Prospects for the commercialization of chemiluminescence-based point-of-care and on-site testing devices. *Anal. Bioanal. Chem.* **2014**, *406*, 5631–5637. [CrossRef]
71. Zhao, S.; Huang, Y.; Shi, M.; Liu, R.; Liu, Y.-M. Chemiluminescence Resonance Energy Transfer-Based Detection for Microchip Electrophoresis. *Anal. Chem.* **2010**, *82*, 2036–2041. [CrossRef]
72. Kenry, G.A.; Lim, C.T. Paper-based MoS2 nanosheet-mediated FRET aptasensor for rapid malaria diagnosis. *Sci. Rep.* **2017**, *7*, 1–8. [CrossRef]
73. Arts, R.R.; Hartog, I.D.; Zijlema, S.E.; Thijssen, V.V.; Van Der Beelen, S.H.E.; Merkx, M.M. Detection of Antibodies in Blood Plasma Using Bioluminescent Sensor Proteins and a Smartphone. *Anal. Chem.* **2016**, *88*, 4525–4532. [CrossRef] [PubMed]
74. Tenda, K.; Van Gerven, B.; Arts, R.; Hiruta, Y.; Merkx, M.; Citterio, D. Paper-Based Antibody Detection Devices Using Bioluminescent BRET-Switching Sensor Proteins. *Angew. Chem. Int. Ed.* **2018**, *57*, 15369–15373. [CrossRef] [PubMed]
75. Hall, M.P.; Unch, J.; Binkowski, B.F.; Valley, M.P.; Butler, B.L.; Wood, M.G.; Otto, P.; Zimmerman, K.; Vidugiris, G.; Machleidt, T.; et al. Engineered Luciferase Reporter from a Deep Sea Shrimp Utilizing a Novel Imidazopyrazinone Substrate. *ACS Chem. Biol.* **2012**, *7*, 1848–1857. [CrossRef]
76. Laver, W.; Air, G.M.; Webster, R.G.; Smith-Gill, S.J. Epitopes on protein antigens: Misconceptions and realities. *Cell* **1990**, *61*, 553–556. [CrossRef]
77. Arts, R.R.; Ludwig, S.K.J.; Van Gerven, B.C.B.; Estirado, E.M.; Milroy, L.L.-G.; Merkx, M. Semisynthetic Bioluminescent Sensor Proteins for Direct Detection of Antibodies and Small Molecules in Solution. *ACS Sens.* **2017**, *2*, 1730–1736. [CrossRef] [PubMed]
78. Rosmalen, M.; Ni, Y.; Vervoort, D.F.M.; Arts, R.; Ludwig, S.K.J.; Merkx, M. Dual-Color Bioluminescent Sensor Proteins for Therapeutic Drug Monitoring of Antitumor Antibodies. *Anal. Chem.* **2018**, *90*, 3592–3599. [CrossRef]
79. Griss, R.; Schena, A.; Reymond, L.; Patiny, L.; Werner, D.; Tinberg, C.E.; Baker, D.; Johnsson, K. Bioluminescent sensor proteins for point-of-care therapeutic drug monitoring. *Nat. Chem. Biol.* **2014**, *10*, 598–603. [CrossRef] [PubMed]
80. Yu, Q.; Xue, L.; Hiblot, J.; Griss, R.; Fabritz, S.; Roux, C.; Binz, P.A.; Haas, D.; Okun, J.G.; Johnsson, K. Semisynthetic sensor proteins enable metabolic assays at the point of care. *Science* **2018**, *361*, 1122–1126. [CrossRef]
81. Xue, L.; Yu, Q.; Griss, R.; Schena, A.; Johnsson, K. Bioluminescent Antibodies for Point-of-Care Diagnostics. *Angew. Chem. Int. Ed.* **2017**, *56*, 7112–7116. [CrossRef]
82. Li, Y.; Zhou, L.; Ni, W.; Luo, Q.; Zhu, C.; Wu, Y.-H. Portable and Field-Ready Detection of Circulating MicroRNAs with Paper-Based Bioluminescent Sensing and Isothermal Amplification. *Anal. Chem.* **2019**, *91*, 14838–14841. [CrossRef]
83. Petryayeva, E.; Algar, W.R. Proteolytic Assays on Quantum-Dot-Modified Paper Substrates Using Simple Optical Readout Platforms. *Anal. Chem.* **2013**, *85*, 8817–8825. [CrossRef]
84. Das, P.; Krull, U.J. Detection of a cancer biomarker protein on modified cellulose paper by fluorescence using aptamer-linked quantum dots. *Analyst* **2017**, *142*, 3132–3135. [CrossRef]
85. Shahmuradyan, A.; Moazami-Goudarzi, M.; Kitazume, F.; Espie, G.S.; Krull, U.J. Paper-based platform for detection by hybridization using intrinsically labeled fluorescent oligonucleotide probes on quantum dots. *Analyst* **2019**, *144*, 1223–1229. [CrossRef] [PubMed]
86. Malhotra, K.; Noor, M.A.F.; Krull, U.J. Detection of cystic fibrosis transmembrane conductance regulator ΔF508 gene mutation using a paper-based nucleic acid hybridization assay and a smartphone camera. *Analyst* **2018**, *143*, 3049–3058. [CrossRef] [PubMed]
87. Noor, M.O.; Shahmuradyan, A.; Krull, U.J. Paper-Based Solid-Phase Nucleic Acid Hybridization Assay Using Immobilized Quantum Dots as Donors in Fluorescence Resonance Energy Transfer. *Anal. Chem.* **2013**, *85*, 1860–1867. [CrossRef] [PubMed]
88. Umrao, S.; Anusha, S.; Jain, V.; Chakraborty, B.; Roy, R. Smartphone-based kanamycin sensing with ratiometric FRET. *RSC Adv.* **2019**, *9*, 6143–6151. [CrossRef]

89. Wang, X.; Wang, S.; Huang, K.; Liu, Z.; Gao, Y.; Zeng, W. A ratiometric upconversion nanosensor for visualized point-of-care assay of organophosphonate nerve agent. *Sens. Actuators B Chem.* **2017**, *241*, 1188–1193. [CrossRef]
90. Hayes, J.; Peruzzi, P.P.; Lawler, S. MicroRNAs in cancer: Biomarkers, functions and therapy. *Trends Mol. Med.* **2014**, *20*, 460–469. [CrossRef] [PubMed]
91. Noor, M.O.; Krull, U.J. Paper-Based Solid-Phase Multiplexed Nucleic Acid Hybridization Assay with Tunable Dynamic Range Using Immobilized Quantum Dots as Donors in Fluorescence Resonance Energy Transfer. *Anal. Chem.* **2013**, *85*, 7502–7511. [CrossRef]
92. Noor, M.O.; Krull, U.J. Camera-Based Ratiometric Fluorescence Transduction of Nucleic Acid Hybridization with Reagentless Signal Amplification on a Paper-Based Platform Using Immobilized Quantum Dots as Donors. *Anal. Chem.* **2014**, *86*, 10331–10339. [CrossRef]
93. Tavares, A.J.; Noor, M.O.; Vannoy, C.H.; Algar, W.R.; Krull, U.J. On-Chip Transduction of Nucleic Acid Hybridization Using Spatial Profiles of Immobilized Quantum Dots and Fluorescence Resonance Energy Transfer. *Anal. Chem.* **2011**, *84*, 312–319. [CrossRef]
94. Zhou, F.; Noor, M.A.F.; Krull, U.J. Luminescence Resonance Energy Transfer-Based Nucleic Acid Hybridization Assay on Cellulose Paper with Upconverting Phosphor as Donors. *Anal. Chem.* **2014**, *86*, 2719–2726. [CrossRef]
95. Doughan, S.; Uddayasankar, U.; Peri, A.; Krull, U.J. A paper-based multiplexed resonance energy transfer nucleic acid hybridization assay using a single form of upconversion nanoparticle as donor and three quantum dots as acceptors. *Anal. Chim. Acta* **2017**, *962*, 88–96. [CrossRef] [PubMed]
96. Bazin, H.; Préaudat, M.; Trinquet, E.; Mathis, G. Homogeneous time resolved fluorescence resonance energy transfer using rare earth cryptates as a tool for probing molecular interactions in biology. *Spectrochim. Acta Part A Mol. Biomol. Spectrosc.* **2001**, *57*, 2197–2211. [CrossRef]
97. Procise Diagnostics. Available online: https://www.procisediagnostics.com/technology (accessed on 21 July 2020).
98. Medical Product Outsourcing Magazine. Available online: https://www.mpo-mag.com/contents/view_breaking-news/2020-02-18/ce-mark-granted-for-point-of-care-inflammatory-marker-assay/ (accessed on 31 July 2020).
99. Perkin Elmer. Available online: https://www.perkinelmer.com/de/category/lance-tr-fret (accessed on 31 July 2020).
100. Hepojoki, S.; Hepojoki, J.; Hedman, K.; Vapalahti, O.; Vaheri, A. Rapid Homogeneous Immunoassay Based on Time-Resolved Förster Resonance Energy Transfer for Serodiagnosis of Acute Hantavirus Infection. *J. Clin. Microbiol.* **2015**, *53*, 636–640. [CrossRef] [PubMed]
101. Rusanen, J.; Toivonen, A.; Hepojoki, J.; Hepojoki, S.; Arikoski, P.; Heikkinen, M.; Vaarala, O.; Ilonen, J.; Hedman, K. LFRET, a novel rapid assay for anti-tissue transglutaminase antibody detection. *PLoS ONE* **2019**, *14*, e0225851. [CrossRef] [PubMed]
102. Akerström, B.; Björck, L. Protein L: An immunoglobulin light chain-binding bacterial protein. Characterization of binding and physicochemical properties. *J. Biol. Chem.* **1989**, *264*, 19740–19746. [CrossRef]
103. Hepojoki, S.; Nurmi, V.; Vaheri, A.; Hedman, K.; Vapalahti, O.; Hepojoki, J. A protein L-based immunodiagnostic approach utilizing time-resolved Förster resonance energy transfer. *PLoS ONE* **2014**, *9*, e106432. [CrossRef]
104. Li, H.; Fang, X.; Cao, H.; Kong, J. Paper-based fluorescence resonance energy transfer assay for directly detecting nucleic acids and proteins. *Biosens. Bioelectron.* **2016**, *80*, 79–83. [CrossRef]

Publisher's Note: MDPI stays neutral with regard to jurisdictional claims in published maps and institutional affiliations.

© 2021 by the authors. Licensee MDPI, Basel, Switzerland. This article is an open access article distributed under the terms and conditions of the Creative Commons Attribution (CC BY) license (http://creativecommons.org/licenses/by/4.0/).

Article

Advanced RuO$_2$ Thin Films for pH Sensing Application

Xinyue Yao [1], Mikko Vepsäläinen [2], Fabio Isa [3], Phil Martin [3], Paul Munroe [1] and Avi Bendavid [1,3,*]

- [1] School of Materials Science and Engineering, University of New South Wales, Kensington, NSW 2052, Australia; Xinyue.yao1@unswalumni.com (X.Y.); p.munroe@unsw.edu.au (P.M.)
- [2] CSIRO Mineral Resources, P.O. Box 312, Clayton South, VIC 3169, Australia; mikko.vepsalainen@csiro.au
- [3] CSIRO Manufacturing, P.O. Box 218, 36 Bradfield Road, Lindfield, NSW 2070, Australia; fabio_Isa@outlook.com (F.I.); phil.martin@csiro.au (P.M.)
- * Correspondence: avi.bendavid@csiro.au

Received: 25 September 2020; Accepted: 9 November 2020; Published: 11 November 2020

Abstract: RuO$_2$ thin films were prepared using magnetron sputtering under different deposition conditions, including direct current (DC) and radio frequency (RF) discharges, metallic/oxide cathodes, different substrate temperatures, pressures, and deposition times. The surface morphology, residual stress, composition, crystal structure, mechanical properties, and pH performances of these RuO$_2$ thin films were investigated. The RuO$_2$ thin films RF sputtered from a metallic cathode at 250 °C exhibited good pH sensitivity of 56.35 mV/pH. However, these films were rougher, less dense, and relatively softer. However, the DC sputtered RuO$_2$ thin film prepared from an oxide cathode at 250 °C exhibited a pH sensitivity of 57.37 mV/pH with a smoother surface, denser microstructure and higher hardness. The thin film RF sputtered from the metallic cathode exhibited better pH response than those RF sputtered from the oxide cathode due to the higher percentage of the RuO$_3$ phase present in this film.

Keywords: ruthenium dioxide; magnetron sputtering conditions; thin film characterisation; pH performance

1. Introduction

The sensing of pH is very important in several chemical and biological processes, such as for applications in water and food quality monitoring, and wearable systems for chronic diseases [1]. In order to optimise the desired response and to eliminate unwanted reactions, pH measurement, and control are both required in many applications, such as blood monitoring, environmental monitoring, water quality monitoring, and various clinical tests. The glass electrode sensor has been commonly used for pH measurement due to its high accuracy, fast response, and ideal Nernst behaviour. However, with the increasing requirements for different applications, glass electrodes exhibit drawbacks such as instability in high temperature environments, poor mechanical properties in high pressure environments, and difficulties in miniaturisation [1]. Therefore, many other pH measurement techniques have been developed recently, one of which is the solid-state metal oxide thin film sensor, based on oxides such as PtO$_2$, IrO$_2$, RuO$_2$, SnO$_2$, and Ta$_2$O$_5$, that has been demonstrated to exhibit excellent pH sensing performance at high temperatures and pressures and these are promising candidates for a future generation of pH sensors. Among these metal oxide materials, magnetron sputtered RuO$_2$ thin films show outstanding properties with near Nernstian pH sensitivity, high conductivity, and excellent mechanical strength and corrosion resistance [2–5], as such it has been researched extensively.

Magnetron sputtering is one of the most attractive physical vapor deposition processes (PVD) for an extensive range of metal oxide materials due to its outstanding advantages, such as high deposition rate, excellent reproducibility, high density, and good quality of deposited thin films [5]. Many researchers have reported excellent pH response for radio frequency (RF) sputtered RuO_2 thin films [6–8]. There have been many studies on the effects of different sputtering conditions on the properties and pH performance of RF sputtered RuO_2 thin films, such as sputtering temperatures in the range of room temperature to 500 °C [9,10], deposition time/thin film thickness [11], and Ar/O ratios ranging between 10 and 2.3 [12,13]. However, there are few reports comparing the differences between direct current (DC) and RF sputtering of RuO_2 thin films and their characteristics and pH response. Furthermore, the effects of different cathodes (either metallic or metal oxide) have not been investigated extensively.

In this study, RuO_2 thin films were sputtered under different sputtering conditions, including DC/RF discharges, metallic/oxide cathodes (Ru/RuO_2), different substrate temperatures (100 °C, 150 °C and 250 °C), sputtering pressures (1.0 Pa and 2.0 Pa), and thicknesses (~200 nm and ~600 nm). The structure and properties of the deposited RuO_2 thin films studied include surface morphology, residual stress, crystal structure, composition, hardness, and elastic modulus. In addition, electrochemical experiments were conducted in terms of pH sensitivity and pH stability. The effect of different sputtering conditions on the RuO_2 thin film properties, especially the effect of different discharges and different cathodes, were studied. Furthermore, the effect of the presence of RuO_3 in these RuO_2 thin films and the correlation between the pH response and—structural characteristics are discussed.

2. Materials and Methods

2.1. Thin Film Fabrication

RuO_2 thin films were deposited using a magnetron sputtering system equipped with an axial turret magnetron head and power supply (AJA DCXS-750, Scituate, MA, USA). The turret head was mounted vertically in the bottom of the vacuum system. The target materials used for sputtering were metallic (Ru) cathode and oxide (RuO_2) cathodes of high purity (>99.0%). The nominal size of the sputter targets was 50 mm in diameter and the distance between the cathodes and the substrate was set at 60 mm. The films were deposited onto (100) conducting silicon wafers with a resistivity of 0.05 Ω-cm. The substrates size was 25.0 mm × 25.0 mm. The substrates were electrically grounded. The deposition system was equipped with rotary and cryogenic pumps and a controlled gas introduction system. A base pressure of 1×10^{-4} Pa was attained in the chamber before the deposition. The oxygen reactive gas and argon inert gas were introduced into the chamber depending on the sputtering target used. The deposition pressure could be set independently of the gas flow by adjusting a throttle valve. The RF or DC powers were set at 100 W or 125 W, respectively. The deposition times varied between 15 to 40 min. The film thickness variation across the substrate was in order of 10.0%. Three groups of RuO_2 thin films were prepared as outlined in Table 1.

Table 1. Deposition conditions for all samples, direct current (DC) and radio frequency (RF)

Sample	S1	S2	S3	D1	D2	D3	D4	T1	T2
DC/RF	RF	RF	RF	DC	DC	RF	RF	DC	DC
Pressure (Pa)	2	2	2	2	2	2	1	2	2
Power (W)	125	125	125	100	100	100	100	100	100
Deposition time (min)	20	20	20	15	15	20	20	40	40
Temperature (°C)	100	150	250	R.T.	250	250	250	250	150
Cathode target	Ru	Ru	Ru	RuO_2	RuO_2	RuO_2	RuO_2	RuO_2	RuO_2

- Samples S1–S3 were deposited by reactive RF sputtering from a metallic cathode target (Ru) with a fixed oxygen partial pressure (Ar/O$_2$ ratio of 4/1), total pressure (2.0 Pa), RF power (125 W), and deposition time (20 min). The substrate temperature was varied from 100 to 250 °C.
- Samples D1–D4 were sputtered from an oxide cathode target (RuO$_2$). D1 and D2 were deposited by DC sputtering with fixed total pressure (2.0 Pa), DC power (100 W), and deposition time (15 min). The substrate temperatures were room temperature and 250 °C respectively. D3 and D4 were deposited by rf sputtering with fixed substrate temperature (250 °C), RF power (100 W) and deposition time (20 min). The deposition pressures were 2.0 and 1.0 Pa, respectively.
- Thicker samples, T1 and T2, were deposited by DC sputtering from an oxide cathode target (RuO$_2$) with a fixed total pressure (2.0 Pa), DC power (100 W) and longer deposition time (40 min). The substrate temperatures were 150 and 250 °C, respectively.

2.2. Film Characterisation

In order to measure the surface morphology of the thin films, a Bruker SPM ICON atomic force microscope (AFM) was employed to determine the roughness and the grain size. The residual stress in the thin films was calculated from the bending height that was measured by a Dektak 3030 surface profilometer and the thickness was determined by scanning electron microscopy (SEM) (ZEISS AURIGA) from the cross-section of the films.

X-ray photoelectron spectroscopy (XPS) measurements of all samples were performed in a SPECS SAGE 150 XPS System using Mg Kα radiation at 10 kV and 10 mA. The system operated at 100 W. The base pressure in the sample chamber of the spectrometer was $<1 \times 10^{-7}$ mbar. The pass energy was 30 eV with a step size of 0.5 eV for broad scanning and was 20 eV with a step size of 0.1 eV for high resolution scanning. The instrumental resolution was 1.3 eV as measured from the FWHM of the $4f_{7/2}$ line for Au at 84.0 eV. Curve fitting of the high-resolution scans and peak area calculations were carried out using Casa XPS software.

Raman scattering measurements were made using a confocal RENISHAW inVia instrument in back-scattering geometry. A solid-state laser with a wavelength of 514 nm was used as the excitation source. The laser power was 1.4 mW and the laser beam was normal to the sample surface. The laser light was focused to a spot size of about 700 nm in diameter onto the sample with an optical microscope. An exposure time of 20 s was used. The signal was detected by a charge coupled device camera and a 2400 lines/mm monochromator. The resolution of the system was about 1 cm^{-1}.

The crystal structure of the RuO$_2$ thin films was characterised by conventional $\theta-2\theta$ X-ray diffraction using an Empyrean XRD Diffractometer with a Cu Kα ($\lambda = 1.5406$ A) source. The hardness and elastic modulus of the thin films were determined by nanoindentation tests with a Hysitron Triboindenter TI 900 using a standard Berkovich indenter. The loading force was set as 2000 µN for all samples. The pH sensitivity and stability of the RuO$_2$ thin films was measured using the open circuit potential (OCP) method in a commercial pH buffer solution (Merck) of different pH values (pH = 2, 4, 7, 10) with a large input resistance.

3. Results and Discussion

3.1. Surface Morphology

The AFM analysis of samples in the S group and D group are shown in Figures 1 and 2. The roughness values, R_a, for these samples are listed in Table 2. The roughness is assumed to increase as the grains tend to coarsen with increasing substrate temperature. From Figure 1b, the sample deposited at 150 °C shows deep voids and protruding grains which leads to a higher roughness than that for sample S3. However, based on Figure 1c, the thin film deposited at 250 °C is denser than that deposited at 150 °C, which is expected in accordance with the higher temperature. The likely reason behind this is that increasing the substrate temperature significantly enhances the lateral mobility of the condensing sputtered target atoms. Temperature-activated incoming depositing atoms tend to fill up the voids instead of self-shadowing, which in turn densifies the film material.

Figure 1. Atomic force microscope (AFM) images of the S group samples; (**a**) S1—100 °C; (**b**) S2—150 °C; (**c**) S3—250 °C.

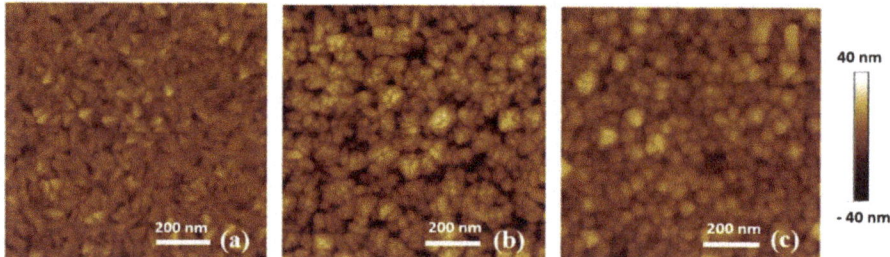

Figure 2. AFM images of D group samples; (**a**) D2, (**b**) D3—and (**c**) D4.

Table 2. Surface roughness of the RuO_2 thin films.

No.	S1	S2	S3	D1	D2	D3	D4	T1	T2
R_a (nm)	12.9	27.1	15.4	3.44	4.61	7.72	5.68	5.20	2.77

In the sample D group, when comparing the AFM images for samples D2 and D3 (Figure 2a,b), the RF sputtered thin film has a rougher surface and larger grain size than the DC sputtered thin film. This is due to the difference in the sputtering plasma. The applied potential will vary significantly between RF and DC sputter deposition. One of the major influences on the film morphology will be the bombardment of the film by high energy (charged) species. The ions within the plasma have higher energy during RF sputtering compared to DC sputtering, which is beneficial for grain growth during deposition [14]. The grain coarsening leads to a higher surface roughness. Samples D3 and D4 in Figure 2b,c indicate that the RuO_2 thin film has a smoother surface and smaller grain size when deposited at a lower sputtering pressure. Similar behaviour was also observed in the case of DC sputtered iridium oxide films deposited onto Si substrates [15]. Samples in the T group (data not shown for brevity) exhibit a relatively low roughness as they were DC sputtered.

There is a significant difference in the roughness between S group sputtered from a metallic cathode (Figure 1c) and D group sputtered from an oxide cathode (Figure 2a). This is because thin films sputtered from the metallic cathode were in an oxygen environment. The oxidation reaction occurs near the Si substrate.

3.2. XPS Analysis

All peaks in the spectra were charge corrected with respect to the C 1s peak (284.6 eV). The Ru doublet peak (Ru $3d_{3/2}$ and Ru $3d_{5/2}$) and Ru 3p peaks were identified in the wide spectral scan. Since the adventitious C 1s peak at 284.6 eV coincides with the Ru $3d_{3/2}$ peak, the use of the Ru 3d peaks for qualitative and quantitative analysis is not reliable, but there have been a few studies using the assignment of Ru 3p peaks for ruthenium oxides, especially for RuO_3 peaks [16]. Thus, fitting of the Ru 3d peaks is used for identification, while the Ru $3p_{3/2}$ peak fitting is used for ratio calculations. Figure 3 shows the peak fitted spectrum for both O 1s and Ru 3d for sample S1.

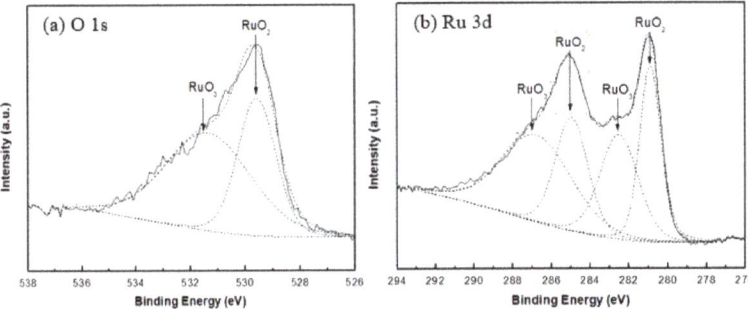

Figure 3. (a) O 1s and (b) Ru 3d peaks region for sample S1.

From the peak fitting analysis for both the O 1s and Ru 3d peaks of sample S1 (Figure 3), two types of ruthenium oxides are identified that contribute to the structure of the deposited thin films. The peak fittings of all samples are similar, indicating that all the samples consist of two different oxide species. According to data from the published literature presented in Table 3, the reported values of RuO_2 and RuO_3 match the binding energies of the Ru 3d and O 1s peaks from the present XPS measurements for all samples. In this case, the ruthenium oxides present in these thin films are identified as RuO_2 and RuO_3.

Table 3. XPS binding energies of different RuO_x compounds from the literature.

RuO_x	O 1s (eV)	Ru $3d_{5/2}$ (eV)	Ru $3d_{3/2}$ (eV)	Ru $3p_{3/2}$ (eV)	Ref.
RuO_2	528.9–529.4	280.1–281.3	284.8–285.0	462.2	[16–20]
RuO_3	530.7–531.2	281.7–282.5	286.6–287.0		[16–20]
RuO_4		282.6–283.3			[16,17,19]

The peak fitting for Ru $3p_{3/2}$ has approximately the same FWHM for both RuO_2 and RuO_3 (3.4 eV). The results of RuO_2/RuO_3 ratios of all samples are listed in Table 4. The results show that the RuO_2 is the dominant oxide species in the thin films deposited at the lower substrate temperatures and especially in the non-oxygen sputtering environment. An obvious trend can be observed according to the samples in the S group that were reactive RF sputtered in an O_2 environment, that is, the percentage of RuO_3 increases as the substrate temperature increases.

Table 4. RuO_2/RuO_3 ratios for all samples.

No.	S1	S2	S3	D1	D2	D3	D4	T1	T2
RuO_2/RuO_3 ratios	1.74	2.61	3.63	4.58	3.78	3.80	3.02	2.46	3.09

3.3. Residual Stress and Raman Spectroscopy

The results of the residual stresses in the thin films were calculated using the Stoney's formula [20], which are listed in Table 5. The residual stress normally increases with increasing substrate temperature as the thin film becomes denser. In the S group, the compressive residual stress increases at first, but then decreases and converts to tensile stress with a further increase in temperature. The higher compressive residual stresses at higher substrate temperatures may be related to the denser microstructure that formed. However, greater grain growth at higher temperatures may also contribute to stress relaxation [21]. In addition, the tensile stress can linearly increase from the thermal mismatch between the thin film and the substrate with increasing deposition temperature [21]. In the case of sample S3, that was deposited at the highest temperature, grains coarsen, and the thermal tensile stress may exceed the intrinsic compressive stress, which resulted in a tensile residual stress in this thin film. The effect of sputtering pressure on the residual stress can also be observed from samples D3 and D4. At a lower sputtering pressure of 1.0 Pa, the compressive residual stress is higher in this thin film. The reason is that the lower sputtering pressure can give a lower sputtering rate, which is beneficial to deposit a denser thin film.

Table 5. Residual stress for all samples.

No.	S1	S2	S3	D1	D2	D3	D4	T1	T2
σ (GPa)	0.25	0.53	−0.30	0.50	0.77	0.34	0.82	0.40	0.41

The Raman spectra for samples in groups S and D are shown in Figure 4. Three Raman-active modes, e.g., A1g and B2g can be observed in the Raman spectra and analysis of spectra for these films can provide insight into the phase constitution of each sample. The B1g mode is too weak to be observed. For the thinner films, there is a combination of RuO_2 and Si peaks at the, e.g., mode region for each sample. The possible reason is that the RuO_2 thin film is deposited with a nm-scale thickness onto the Si substrate, so the incident laser penetrates the RuO_2 thin film and interacts with the Si substrate. In this case, peak fitting is applied to specify the position of the, e.g., frequency mode and is also used to assign the peak position of the A1g and B2g modes for RuO_2.

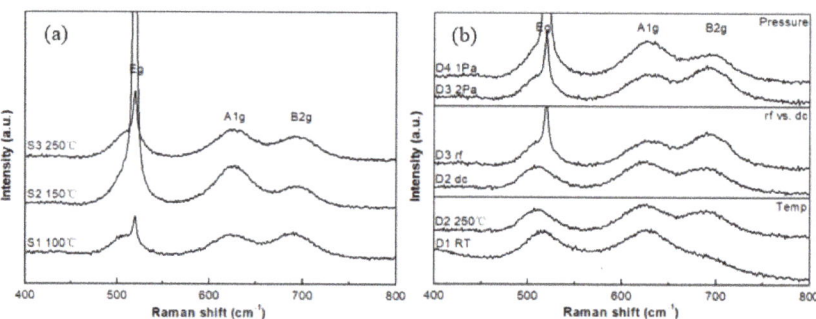

Figure 4. Raman spectra for samples in the (**a**) S group (**b**) D group with deposition pressure, RF and DC sputtering and deposition temperature variations.

The three major Raman-active modes, e.g., A1g and B2g for single-crystal RuO_2 are located at 528, 644, and 716 cm^{-1}, respectively [22,23]. In this experiment, all the three Raman-active modes show a red shift in the peak location. The red shift has a linear relationship with the residual stress according to Meng and Dos Santos [24]. In fact, the stress should have an influence on all modes, but the stress effect on A1g mode is greater than the others [24]. Figure 5 shows the relationship between the A1g peak position and the residual stress of the S and D groups. In the case of the group D samples, films D1, D2, and D3 show that the Raman shift increases with the residual stress linearly. Sample D4 (stress = 0.82 GPa) falls out of the trend as it is the only sample deposited at a lower pressure (1.0 Pa) than the other three samples deposited at 2.0 Pa. The deposition pressure influences the bombardment energy of the depositing atoms resulting in the modifications of the properties of the films such as texture, morphology, composition and stress. The data of the group S show a non-linear relationship. This difference in the trend can be attributed to the surface morphologies and film texture between the two groups of samples. In addition, the increasing percentage of RuO_3 for group S sputtered from the metallic cathode that alters the O/Ru ratio. Parker et al. [25] indicated that the Raman peak positions are dependent on the O/Ti ratio in TiO_2 films.

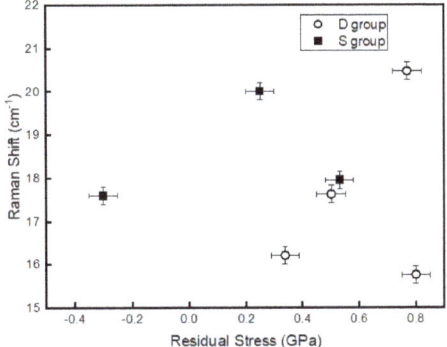

Figure 5. The relationship between the residual stress and Raman shift for the A1g mode.

Chan et al. [26] assigned the peak at 800 cm^{-1} to RuO_3 using surface-enhanced Raman spectroscopy. However, here the Raman results do not indicate a peak at 800 cm^{-1}. This is likely because the region of the composition that XPS measures is only the near surface region of the thin film, while the region of the structure that Raman spectroscopy measures is much deeper into the thin film and hence the substrate (that is, a strong Si peak can be observed in the Raman spectra). The surface region may consist of RuO_2 and RuO_3, while the bulk region of thin film may only consist of RuO_2. In this case, the RuO_2 thin film can be considered as a layered structure. This is in agreement with the work of Chou et al. [8].

3.4. X-ray Diffraction Analysis

From Figure 6, for films deposited (S group) at both 100 and 150 °C, the preferred crystal orientation of RuO_2 thin film is (101). With increasing substrate temperature up to 250 °C, there is a peak at (110) which increases in intensity with temperature. This result is similar to that recorded previously [27–29], where the RuO_2 thin films show a preferred orientation along the (101) over the temperature range from 100 °C to 300 °C. There is a relatively weak (200) peak at the highest temperature of 250 °C, which was reported to increase with the temperature increases to above 300 °C [23]. All observed peaks of the RuO_2 films were assigned to the tetragonal rutile structure with lattice parameters of c = 4.50 Å and c = 3.05 Å.

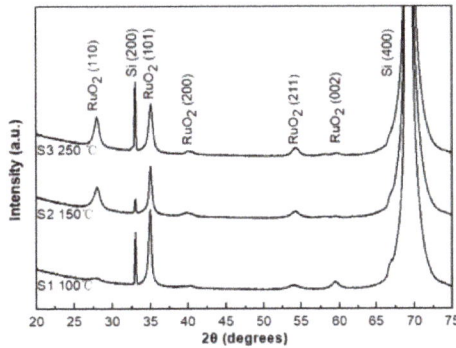

Figure 6. X-ray diffraction patterns of samples from the S group.

For the sample prepared by RF sputtering at the lower pressure of 1.0 Pa, the thin film exhibits relatively weak crystallinity (Figure 7a). From Figure 7b, the thin film prepared by DC sputtering shows a polycrystalline structure when compared to that prepared by RF sputtering. The rf sputtered thin films exhibit the (101) preferential orientation. As the temperature increases to 250 °C, there is a change in preferential orientation from (101) to (110). For the sample prepared by DC sputtering at room temperature, (not shown here) the thin film exhibits a dominant crystal orientation of (101), which is different from the reported results where the RuO$_2$ samples, prepared by RF sputtering, showed an amorphous structure at room temperature [27,28]. The XRD patterned of DC sputtered film deposited at 150 °C (T2) is shown in Figure 7c, exhibiting a dominant crystal orientation of (101), A broad peak was located at around 55° that indicates a short-range periodic arrangement of RuO$_2$ (211). Compared to the samples sputtered from the metallic Ru cathode, which have polycrystalline structure (Figure 6 (S3) and Figure 7b (D3)), the thin films prepared from the RuO$_2$ oxide cathode have a preferential (101) orientation. This is presumably because the oxidation reaction of the metal target atoms at the substrate causes the change in the deposition behaviour of the oxygen species. Compared to the sample prepared under the same condition, but with a thinner film (Figure 7b (D2) and (Figure 7c (T2)), the thicker thin film shows a strong preferential (101) orientation. In the XRD measurements, no peaks of the RuO$_3$ phase were observed for all the samples studied here. As indicated earlier, the RuO$_3$ phase was identified with the XPS technique which is a surface sensitive technique (~5 to 10 nm depth) where the sensitivity depth of the XRD analysis is in the micrometers range. Therefore, due to the different depth sensitivity of these techniques, one may obtain different information if the film is not homogenous with thickness. We deduce that the RuO$_3$ is present only on the outer surface of the films and not in the bulk since no XRD peaks assigned to RuO$_3$ were observed.

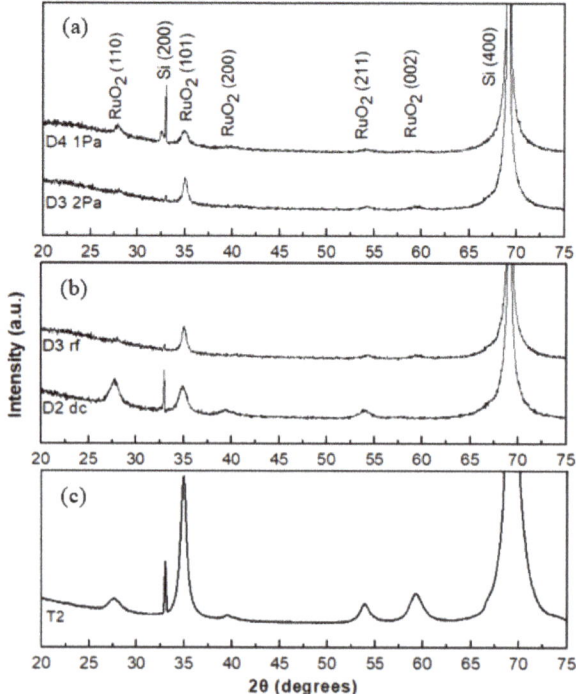

Figure 7. X-ray diffraction patterns of the D group samples (**a**) RF discharge: 1.0 Pa and 2.0 Pa; (**b**) DC/RF difference; and (**c**) T2 with thicker thickness (~700 nm).

3.5. Hardness and Modulus Results

The load-unload curves for samples in the D group and T group all show good homogeneity. Figure 8 shows the load-unload curves for sample D1. The load-displacement (p-h) curves for samples in the S group, however, show a greater variation, which is mainly because of the non-ideal sample surface for these samples. The high roughness and the relatively small thickness may also contribute to this effect. In this case, the deviations of the data are large, and the results are unreliable. The hardness and elastic reduced modulus values for all samples are listed in Table 6.

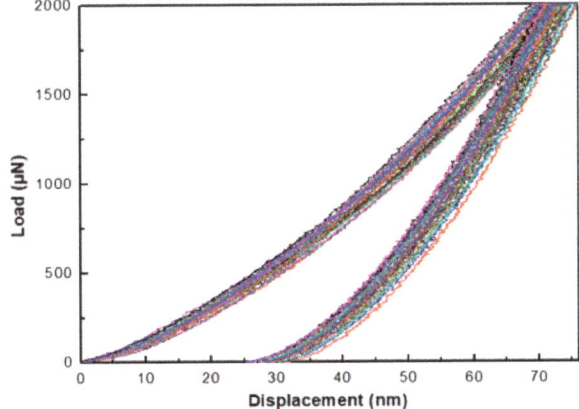

Figure 8. Load-unload curves of sample D1 under 2000 µN maximum loading force.

Table 6. Hardness and elastic reduced modulus for all samples.

Sample	Hardness (GPa)	Elastic Reduced Modulus (GPa)
S1	6.1	144.8
S2	3.8	120.9
S3	6.2	133.9
D1	13.6	164.9
D2	17.2	190.4
D3	10.3	157.8
D4	11.5	156.7
T1	12.3	176.1
T2	12.0	172.7

In brief, however, the hardness increases as the substrate temperature increases because the thin films become denser at high temperature. Residual stress also influences the hardness; compressive stress makes thin films harder, while tensile stress makes thin films softer. Sample D2, that retains the highest compressive residual stress, exhibits the highest hardness. Although the hardness result for sample S3 that is under a tensile residual stress is not reliable, based on its p-h curves, it can still be observed that this thin film exhibits a relatively low hardness. The low hardness of all samples in group S may be attributed to the greater grain size and the presence of RuO_3.

Sample D3, prepared by RF sputtering, is softer than sample D2 prepared by DC sputtering. This is because of the different plasma effect of RF discharges that makes the grains coarsen. Sample D4 deposited at a lower pressure exhibits a higher hardness than sample D3 due to its denser microstructure at a lower sputtering rate. The thin film hardness is highly dependent on the sputtering conditions. Búc et al. [9] reported that a sputtered RuO_2 thin film exhibited a hardness of 9.4 ± 1.7 GPa, while Zhu et al. [29] reported that the thin film exhibited a hardness of 20.4 ± 2.4 GPa. The hardness results obtained in this experiment are in the range of these reported values.

3.6. Electrochemical Results

Figure 9 shows the potential of saturated calomel electrode (SCE) versus pH values for the different groups of samples. The pH sensitivity of all samples is listed in Table 7. Samples S1, S3, T1, and T2 show near-Nernstian slopes of 53.6, 56.4, 57.4, and 54.1 mV/pH, respectively.

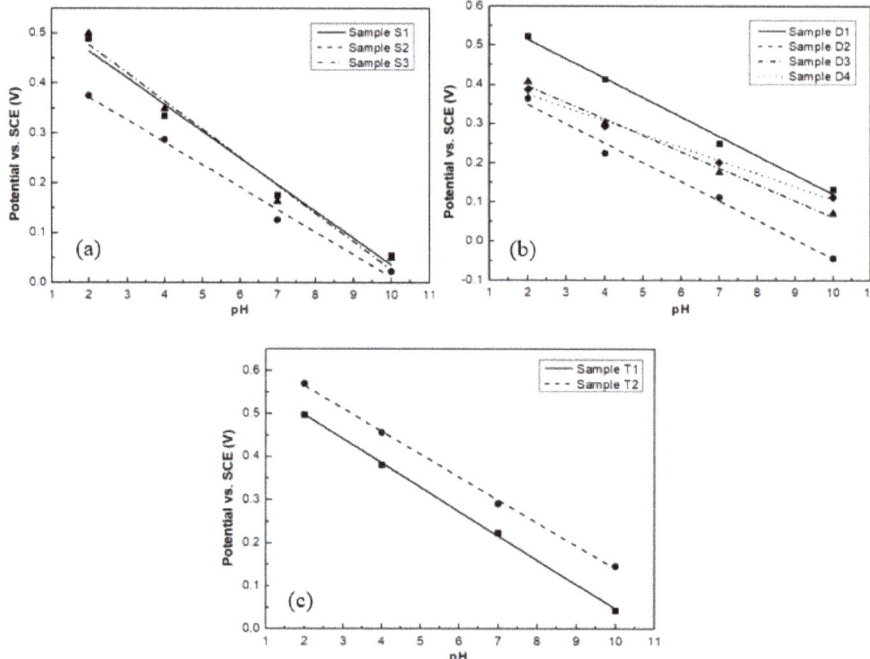

Figure 9. pH sensitivity of samples (**a**) S group; (**b**) D group; (**c**) T group.

Table 7. pH sensitivity and linearity of all samples.

Sample	Sensitivity (mV/pH)	Linearity
S1	53.6	0.9735
S2	45.0	0.9882
S3	56.4	0.9708
D1	49.3	0.9907
D2	49.1	0.9835
D3	41.8	0.9896
D4	33.6	0.9846
T1	57.4	0.9986
T2	54.1	0.9980

The poor pH sensitivity for sample S2 could be attributed to its high porosity. From its AFM image, shown in Figure 1b, the surface exhibits deep voids and protruding grains indicating significant pore formation in the RuO_2 thin film. The enhanced pore formation increases the scattering of the charge carriers and thus reduces the carrier mobility [30,31], which in turn results in the low pH sensitivity of this thin film. Furthermore, sample S2 also exhibits a potential drift. This is attributed to the trapping of hydrogen and hydroxide ions when they diffuse into the pores of the thin film [32].

When comparing the results of the pH sensitivity and the percentage of RuO_3 in the thin film, from samples S1 to S3 and T1 to T2, it can be also seen that RuO_2 thin films with a higher percentage of

RuO$_3$ have a better pH sensitivity. This is because a higher percentage of RuO$_3$ leads to the higher oxygen ratio in the thin film, which increases the redox reaction speed.

In the D group, sputtered from the RuO$_2$ cathode, it can be seen that thin films prepared by DC sputtering have a higher pH sensitivity than thin films prepared by RF sputtering. Both samples D3 and D4 exhibit the lowest pH sensitivity probably due to their smaller thickness, which is in agreement with the published literature [33], where the pH sensitivity of RuO$_2$ electrodes decreases as their thickness decreases. This can also be seen in samples D2 and T1, which were prepared under the same sample conditions, but with different thicknesses. The pH sensitivity of the thicker sample T1 is highest. The poor pH response of the thinner RuO$_2$ films indicates that there is insufficient RuO$_2$ in the coating to react with the solution. This would result in the extensive exposure of the underlying Si substrate, which could form a pH dependent galvanic couple between the substrate material and solution and thus reduce the pH sensing performance.

All the samples show a stable output potential for all pH values over time, except sample S2 that exhibited a potential drift for both acidic and alkaline solutions. Figure 10 shows the potential versus time at different pH buffer solution of sample S3 as a representative example where the stability of the output is clearly evident.

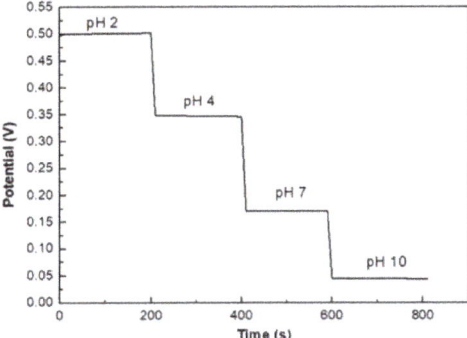

Figure 10. pH stability of sample S3.

4. Conclusions

In this study, the effects of magnetron sputtering conditions, including DC/RF discharge, metallic/oxide cathode target, different substrate temperatures, pressures, and thicknesses, on the characterisation and pH performance of the RuO$_2$ thin film have been investigated.

- The effects of metallic/oxide cathode. The thin films sputtered from the metallic cathode were much rougher than those sputtered from the oxide cathode due to the oxidation reaction near the substrate/as-deposited thin film and the bombardment of negative oxygen ions during sputtering process. This leads to a higher proportion of RuO$_3$ in the thin film. The thin films sputtered from the oxide cathode were found to be much harder.
- The effects of DC/rf discharge. The RuO$_2$ thin films deposited by RF sputtering were rougher than that deposited by DC sputtering. The rf discharge is beneficial to the grain growth in the thin film, which leads to a softer film. The DC sputtered thin films have a higher pH sensitivity response than the RF sputtered thin film.
- The effects of substrate temperature. The RuO$_2$ thin film is rougher and denser at higher substrate temperatures due to greater grain growth. The compressive residual stress increases with increasing temperature. The percentage of RuO$_3$ in the RuO$_2$ thin film increases as the substrate temperature increases.

- The effects of sputtering pressure. The RuO$_2$ thin film are rougher and less dense when deposited at a higher sputtering pressure due to the higher sputtering rate. At a lower pressure, the thin film retains a higher compressive residual stress, which results in a higher hardness. The percentage of RuO$_3$ is higher at lower sputtering pressure.
- The effects of the presence of RuO$_3$. The presence of RuO$_3$ in the RuO$_2$ thin film results in the thin film to be rougher and softer. The higher percentage of RuO$_3$ in the thin film leads to a better pH response. The Raman red shift is related to both the residual stress and the O/Ru ratio.
- Sample S3 RF, sputtered from a metallic cathode at 250 °C, and thick sample T1, DC sputtered from the oxide cathode at 250 °C, have near-Nernstian pH sensitivities of 56.4 and 57.4 mV/pH, respectively. The RuO$_2$ thin film RF sputtered from the metallic cathode at higher temperature exhibits a good pH performance with a thinner thickness. However, the thin film is rougher, less dense, and softer. The RuO$_2$ thin film DC sputtered from the oxide cathode at higher temperature exhibited a good pH performance with a smoother surface, denser microstructure and higher hardness.

Author Contributions: Conceptualisation–resources, A.B.; methodology–writing—original draft preparation, X.Y.; formal analysis, M.V., F.I., and P.M. (Phil Martin); supervisions; P.M. (Paul Munroe) All authors have read and agreed to the published version of the manuscript.

Funding: This research received no external funding.

Conflicts of Interest: The authors declare no conflict of interest.

References

1. Miao, Y.; Chen, J.; Keming, F. New technology for the detection of pH. *J. Biochem. Biophys. Methods* **2005**, *63*, 1–9. [CrossRef]
2. Kreider, K.; Tarlov, M.; Cline, J. Sputtered Thin-Film pH electrodes of Platinum, Palladium, Ruthenium and Iridium Oxides. *Sens. Actuators B Chem.* **1995**, *28*, 167–172. [CrossRef]
3. Kurzweil, P. Metal Oxides and Ion-Exchanging Surfaces as pH Sensors in Liquids: State-of-the-Art and Outlook. *Sensors* **2009**, *2*, 4955–4985. [CrossRef] [PubMed]
4. Xu, B.D.; Zhang, W.D. Modification of vertically aligned carbon nanotubes with RuO$_2$ for a solid-state pH sensor. *Electrochim. Acta* **2010**, *55*, 2859–2864. [CrossRef]
5. Maurya, D.; Sardarinejad, A.; Alameh, K. Recent Developments in R.F. Magnetron Sputtered Thin Films for pH Sensing Applications—An Overview. *Coatings* **2014**, *4*, 756–771. [CrossRef]
6. Maurya, D.; Sardarinejad, A.; Alameh, K. High-sensitivity pH sensor employing a sub-micron ruthenium oxide thin-film in conjunction with a thick reference electrode. *Sens. Actuators A Phys.* **2013**, *203*, 300–303. [CrossRef]
7. Liao, Y.H.; Chou, J.C. Preparation and characteristics of ruthenium dioxide for pH array sensors with real-time measurement system. *Sens. Actuators B Chem.* **2008**, *128*, 603–612. [CrossRef]
8. Chou, J.C.; Liu, S.I.; Chen, S.H. Sensing Characteristics of Ruthenium Films Fabricated by Radio Frequency Sputtering. *Jpn. J. Appl. Phys.* **2005**, *44*, 1403–1408. [CrossRef]
9. Búc, D.; Mikula, M.; Music, D.; Helmersson, U.; Jin, P.; Nakao, S.; Li, K.Y.; Shum, P.W.; Zhou, Z.; Caplovicova, M. Ruthenium oxide films prepared by reactive unbalanced magnetron sputtering. *J. Electr. Eng.* **2004**, *55*, 39–42.
10. Jia, Q.; Shi, Z.; Jiao, K.; Anderson, W.; Collins, F. Reactively sputtered RuO$_2$ thin film resistor with near zero temperature coefficient of resistance. *Thin Solid Film.* **1991**, *196*, 29–34. [CrossRef]
11. Maurya, D.; Sardarinejad, A.; Alameh, K. The effects of sensing electrode thickness on ruthenium oxide thin-film pH sensor. *Sens. Actuators A Phys.* **2014**, *214*, 15–19. [CrossRef]
12. Huang, D.; Chen, J. Material characteristics and electrical property of reactively sputtered RuO2 thin films. *Thin Solid Film* **2001**, *382*, 139–145. [CrossRef]
13. Maurya, D.; Sardarinejad, A.; Alameh, K.; Comini, E. The pH Sensing Properties of RF Sputtered RuO$_2$ Thin-Film Prepared Using Different Ar/O$_2$ Flow Ratio. *Materials* **2015**, *8*, 3352–3363. [CrossRef]
14. Ellmer, K.; Wendt, R. DC and RF (reactive) magnetron sputtering of ZnO:Al films from metallic and ceramic targets: A comparative study. *Surf. Coat. Technol.* **1997**, 21–26. [CrossRef]

15. Negi, S.; Bhandari, R.; Rieth, L.; Solzbacher, F. Effect of sputtering pressure on pulsed-DC sputtered iridium oxide films. *Sens. Actuators B Chem.* **2009**, *137*, 370–378. [CrossRef]
16. Morgan, D.J. Resolving ruthenium: XPS studies of common ruthenium materials. *Surf. Interface Anal.* **2015**, *47*, 1072–1079. [CrossRef]
17. Linford, M.R. *Introduction to Surface and Material Analysis and to Various Analytical Techniques*; Vacuum Technology & Coating: Weston, CT, USA, 2014; pp. 27–32.
18. Bell, W.E.; Tagami, M. High-temperature chemistry of the ruthenium—Oxygen system. *J. Phys. Chem.* **1963**, *67*, 2432–2436. [CrossRef]
19. Hrbek, J.; Van Campen, D.; Malik, I. The early stages of ruthenium oxidation. *J. Vac. Sci. Technol. A* **1995**, *13*, 1409–1412. [CrossRef]
20. Stoney, G.G. The tension of metallic films deposited by electrolysis. *Proc. Royal. Soc. Ser. A* **1909**, *82*, 172. [CrossRef]
21. Hong, S.; Yang, K. Stress measurements of radio-frequency reactively sputtered RuO_2 thin films. *J. Appl. Phys.* **1996**, *880*, 22–826. [CrossRef]
22. Mar, S.; Chen, C.; Huang, Y.; Tiong, K. Characterization of RuO_2 thin films by Raman spectroscopy. *Appl. Surf. Sci.* **1995**, *90*, 497–504. [CrossRef]
23. Meng, L.; Teixeira, V.; Dos Santos, D. Raman spectroscopy analysis of magnetron sputtered RuO_2 thin films. *Thin Solid Film.* **2003**, *442*, 93–97. [CrossRef]
24. Meng, L.; Dos Santos, M. Study of residual stress on RF reactively sputtered RuO_2 thin films. *Thin Solid Film.* **2000**, *375*, 29–32. [CrossRef]
25. Parker, J.; Siegel, R. Raman Microprobe study of nanophase TiO_2 and oxidation-induced spectral changes. *J. Mater. Res.* **1990**, *5*, 1246–1252. [CrossRef]
26. Chan, H.Y.H.; Takoudis, C.G.; Weaver, M.J. High-pressure oxidation of ruthenium as probed by surface-enhanced Raman and X-ray photoelectron spectroscopies. *J. Catal.* **1997**, *172*, 336–345. [CrossRef]
27. Lim, W.T.; Cho, K.R.; Lee, C.H. Structural and electrical properties of RF-sputtered RuO_2 films having different conditions of preparation. *Thin Solid Film.* **1999**, *348*, 56–62. [CrossRef]
28. Reddy, Y.V.; Mergel, K. Structural and electrical properties of RuO_2 thin films prepared by RF-magnetron sputtering and annealing at different temperatures. *J. Mater. Sci.: Mater. Electron.* **2006**, *17*, 1029–1034. [CrossRef]
29. Zhu, J.B.; Yeap, K.B.; Zeng, K.; Lu, L. Nanomechanical characterization of sputtered RuO_2 thin film on silicon substrate for solid state electronic devices. *Thin Solid Film.* **2011**, *519*, 1914–1922. [CrossRef]
30. Liao, Y.H.; Chou, J.C. Potentiometric Multisensor Based on Ruthenium Dioxide Thin Film with a Bluetooth Wireless and Web-Based Remote Measurement System. *IEEE Sens. J.* **2009**, *9*, 1887–1894. [CrossRef]
31. Cui, H.N.; Teixeira, V.; Meng, L.J.; Martins, R.; Fortunato, E. Influence of oxygen/argon pressure ratio on the morphology, optical and electrical properties of ITO thin films deposited at room temperature. *Vacuum* **2008**, *82*, 1507–1511. [CrossRef]
32. Zhuiykov, S. Morphology of Pt-doped nanofabricated RuO_2 sensing electrodes and their properties in water quality monitoring sensors. *Sens. Actuators B Chem.* **2009**, *136*, 248–256. [CrossRef]
33. Lonsdale, W.; Alameh, M.; Wajrak, K. Effect of conditioning protocol, redox species and material thickness on the pH sensitivity and hysteresis of sputtered RuO_2 electrodes. *Sensors. Actuators B Chem.* **2017**, *252*, 251–256. [CrossRef]

Publisher's Note: MDPI stays neutral with regard to jurisdictional claims in published maps and institutional affiliations.

© 2020 by the authors. Licensee MDPI, Basel, Switzerland. This article is an open access article distributed under the terms and conditions of the Creative Commons Attribution (CC BY) license (http://creativecommons.org/licenses/by/4.0/).

Article

Maturity Prediction in Yellow Peach (*Prunus persica* L.) Cultivars Using a Fluorescence Spectrometer

Alessio Scalisi [1,2,*], Daniele Pelliccia [3,4,*] and Mark Glenn O'Connell [1,2]

1. Agriculture Victoria, Tatura, VIC 3616, Australia; mark.oconnell@agriculture.vic.gov.au
2. Food Agility CRC Ltd., Ultimo, NSW 2007, Australia
3. Rubens Technologies Pty Ltd., Rowville, VIC 3178, Australia
4. Instruments & Data Tools Pty Ltd., Rowville, VIC 3178, Australia
* Correspondence: alessio.scalisi@agriculture.vic.gov.au (A.S.); daniel@rubenstech.com (D.P.)

Received: 14 October 2020; Accepted: 15 November 2020; Published: 17 November 2020

Abstract: Technology for rapid, non-invasive and accurate determination of fruit maturity is increasingly sought after in horticultural industries. This study investigated the ability to predict fruit maturity of yellow peach cultivars using a prototype non-destructive fluorescence spectrometer. Collected spectra were analysed to predict flesh firmness (FF), soluble solids concentration (SSC), index of absorbance difference (I_{AD}), skin and flesh colour attributes (i.e., a* and H°) and maturity classes (immature, harvest-ready and mature) in four yellow peach cultivars—'August Flame', 'O'Henry', 'Redhaven' and 'September Sun'. The cultivars provided a diverse range of maturity indices. The fluorescence spectrometer consistently predicted I_{AD} and skin colour in all the cultivars under study with high accuracy (Lin's concordance correlation coefficient > 0.85), whereas flesh colour's estimation was always accurate apart from 'Redhaven'. Except for 'September Sun', good prediction of FF and SSC was observed. Fruit maturity classes were reliably predicted with a high likelihood (*F1*-score = 0.85) when samples from the four cultivars were pooled together. Further studies are needed to assess the performance of the fluorescence spectrometer on other fruit crops. Work is underway to develop a handheld version of the fluorescence spectrometer to improve the utility and adoption by fruit growers, packhouses and supply chain managers.

Keywords: flesh colour; flesh firmness; index of absorbance difference (I_{AD}); machine learning; non-destructive measurements; pigments; sensor; ripeness; skin colour; soluble solids

---

1. Introduction

Fruit maturity indices are used to inform harvest logistics and supply chain management decisions for the delivery of fruit with optimal quality to consumers. Soluble solids concentration (SSC), flesh firmness (FF), starch concentration, titratable acidity, skin and flesh colour, fruit size and shape, ethylene production and respiration rate are useful indices used for stone and pome fruit maturity assessment [1,2]. However, until more recent times, the determination of these parameters has been mostly carried out destructively on small sample sizes, leading to time-consuming operations, expensive labour and often subjective data influenced by operators' skills. Only in recent years, the introduction of spectrometers has led to the increasing adoption of non-destructive devices for food quality estimation (e.g., near-infrared, fluorescence meters, mid-infrared and multispectral/hyperspectral imagery) [3–6]. Based on the maturity index of interest, one technology can be more reliable than others. The handheld, non-destructive Delta Absorbance (DA)meter was introduced by Ziosi et al. in 2008 [7] to determine the index of absorbance difference between 670 and 720 nm (I_{AD}) and has often been used for maturity estimation in stone fruits thereafter [8,9]. The I_{AD} measures chlorophyll a concentration in peach fruit and it has been shown to correlate with ethylene production and respiration rates [7]. Currently, the Australian stone fruit industry recommends the use of FF and I_{AD} for maturity assessment, whilst SSC and fruit size are mostly used as quality parameters. Near-infrared (NIR) spectrometers have been reliably adopted for the estimation of SSC and dry matter in many different fruits—e.g., apple [10–12], pear [13,14], kiwifruit [15] and stone fruits [16–19]—as different wavelengths in the NIR region are very well correlated with the absorbance and reflectance of water and soluble sugars. Prediction of FF in stone fruits via NIR spectrometry is not as reliable as for SSC and dry matter, as this index is influenced by a combination of several physiological and physical factors (e.g., changes in soluble sugars and structural carbohydrates, pectins and physical damage) and does not consistently correlate with specific spectral wavelengths [19]. Other non-destructive technologies such as magnetic resonance, although very precise for the estimation of some maturity indices [20], remain too costly for wide-scale use by industry.

In yellow peaches, skin ground colour and flesh colour have been previously associated with maturity [1,21–25]. Both these parameters are not susceptible to the influence of light. Therefore, during the ripening process, their colour typically turns from green to yellow to orange-red. However, several new yellow peach cultivars have been bred for uniform red skin colouration, that in combination with high SSC represent important key quality parameters for consumers in Australia and south-eastern Asian markets—the most relevant export destinations for Australian stone fruits. This has led to reduced presence or absence of skin ground colour in some cultivars, making skin colour not an ideal candidate for maturity determination in yellow peach cultivars. On the other hand, flesh colour is a more stable attribute among yellow peach cultivars, but its determination requires sample destruction.

Flesh colour is traditionally determined by visual assessments, but colour-measuring devices can be used to determine colour attributes in the CIELAB space [26]. This three-dimensional space is characterised by L* (i.e., a lightness coefficient ranging from 0 (black) to 100 (white)), a* (i.e., a scale of redness to greenness ranging from −60 (true green) to + 60 (true red)) and b* (i.e., a scale of yellowness to blueness ranging from −60 (true blue) to + 60 (true yellow)) [27]. Hue angle (H°) and chroma (C*) colour attributes can be calculated from L*, a* and b* [28]. The H° is calculated as the arc tangent of b*/a* and represents a 360° wheel where 0 or 360° is true red, 90° is true yellow, 180° is true green and 270° is true blue; C* represents colour saturation or "vividness"—diluted with white or darkened with black, with more positive values being brighter and more negative values being duller and is calculated as the square root of $(a^*)^2 + (b^*)^2$ [29]. Flesh a* has been associated with maturity in clingstone peaches [21,22] and, more specifically, to carotenoid content, the most characteristic pigment in the pulp of yellow peach [24]. However, being on a scale from green to red, a* does not take into consideration the yellow colour component—mostly given by carotenoids—that is instead part of H°. Therefore, flesh H° has also been linked to maturity in yellow peach [30] and apricot [31]. In mature yellow peaches, flesh tends to be more orange or red than immature fruit, with a difference of approximately 5°H [30]. Slaugther et al. [30] also observed that flesh H° is a better indicator of maturity than flesh firmness in yellow clingstone peaches. Skin or flesh H° in yellow peaches was predicted using a Vis/NIR interactance spectrometer [32] and a portable fluorometer [26]. Fluorescence spectroscopy has been linked to the concentration of pigments such as chlorophylls, anthocyanins, flavonols and carotenoids in fruit tissues [33–35], with these pigments showing reflectance and absorbance features in the 400–800 nm spectral range [36,37].

The general aim of this study was to assess the ability to non-destructively predict the maturity of yellow peach cultivars using a fluorescence spectrometer. More specifically, the accuracy of the prediction of well-established and less-adopted maturity indices expressed in continuous variables such as SSC, FF, I_{AD}, skin and flesh colour attributes, and three maturity classes expressed in a categorical variable (i.e., immature, harvest-ready and mature) were investigated. The hypothesis was that multivariate analyses of pigment fluorescence spectra may lead to accurate estimations of maturity (>75%).

2. Materials and Methods

2.1. Experimental Site, Plant Material and Fruit Sampling

The experiment was carried out at the Tatura SmartFarm, Agriculture Victoria, Australia (36°26′7″ S and 145°16′8″ E, 113 m a.s.l.) between January and March 2020 on 5–7 years old trees of four yellow peach cultivars (*Prunus persica* L. Batsch, 'August Flame', late 'O'Henry', 'Redhaven' and 'September Sun'). The experimental orchard's soil had clay-loam texture and trees were irrigated, fertigated, pruned and pest/disease managed following established local commercial practices. Fruit of the four cultivars were harvested based on the I_{AD} maturity classes reported by the Victorian Horticulture Industry Network [38]. A DA-meter (TR Turoni, Forlì, Italy) was used to determine the I_{AD} and fruit were harvested at the harvest-ready (onset of ethylene) maturity stage [38], with the exception of 'Redhaven' fruit that were harvested late (i.e., mature—peak of ethylene emission) due to temporary unavailability of the DA-meter. Two hundred fruit per cultivar were picked in mid-January 2020 ('Redhaven'), mid-February 2020 ('O'Henry' and 'August Flame') and early-March 2020 ('September Sun'). Fruit were hand-picked early in the morning (i.e., before 0800 h, AEST) from different orchard rows and trees, and with a sampling strategy of selecting a wide range of fruit size, shape and skin red colour coverage within each cultivar. After harvest, fruit were brought to the laboratory and left to adjust to a constant temperature of 25 °C for approximately two hours. Meanwhile, one cheek per fruit was marked and specimens were numbered progressively from 1 to 200. All the fruit non-destructive and destructive measurements were carried out within 5–8 h of harvest.

2.2. Fluorescence Spectroscopy

Fluorescence spectra were collected on the marked cheek of each fruit using a custom-built fluorescence spectrometer prototype (Rubens Technologies Pty Ltd., Rowville, VIC, Australia) featuring excitation UV LED sources emitting at a wavelength of 400 nm. The spectral sensitivity of the device was in the range 350–950 nm. Since the fluorescence emission is orders of magnitude weaker than the excitation UV intensity, a high-pass filter with cut-off wavelength of 500 nm was placed at the spectrometer entrance to cut the primary UV light and retain only the fluorescent emission from the sample. Recorded spectra represented the average of five readings per fruit obtained in 5–10 s. Data were collected using a custom software running on the Windows 10 operating system and via a USB connection to a laptop.

Spectral measurements were carried out inside a black enclosure to remove influence from external light. The distance between the spectrometer entrance window and the surface of the fruit was kept constant at a value of 80 mm. The measured spectra were normalised to the total emission.

2.3. Determination of Maturity Indices and Maturity Classes

Fruit equatorial diameter from cheek to cheek (FD, mm) and fruit weight (FW, g) were obtained using a digital calliper and a digital scale with two decimal places, respectively. Subsequently, the marked cheek of each individual fruit previously exposed to the fluorescence spectrometer was scanned with the DA-meter to obtain the I_{AD} and data were downloaded into a PC using an internal micro-SD card. Fruit were classified into three maturity classes based on cultivar-specific I_{AD} thresholds obtained from relationships with fruit ethylene emission [38], as shown in Table 1.

Skin colour was determined on a single point per fruit—in the centre of the marked cheek—using a portable spectrophotometer (Nix Pro™, Nix Sensor Ltd., Hamilton, ON, Canada) with a 14 mm aperture, a D50 illuminant and a 2° observer angle. The Nix Pro™ spectrophotometer was operated using the Nix Pro™ Color Sensor application on an Android smartphone via Bluetooth connectivity. Data were logged in the smartphone memory and downloaded at the end of each measurement session. Colour data was stored in the RGB, XYZ, CMYK, HEX and CIELAB formats, but only the latter was used in this study (i.e., L*, a*, b*, C* and H°). Next, fruit skin was removed from the marked cheek using a potato peeler and the flesh was immediately scanned for colour assessment using the Nix Pro™ spectrophotometer with the same methodology adopted for skin colour determination.

The portion of fruit flesh that was scanned with the Nix Pro™ was then exposed to a penetrometer (FT327, FACCHINI srl, Alfonsine, Italy) with an 8 mm tip and FF was measured on a scale from 0 to 15 kgf. Lastly, a few drops of juice (~2 mL) were extracted from the same area and measured with a digital refractometer (PR-1; ATAGO CO., LTD., Saitama, Japan) to obtain SSC (°Brix).

Table 1. Fruit maturity classes for yellow-flesh peach based on cultivar-specific thresholds of index of absorbance difference (I_{AD}) according to the Victorian Horticulture Industry Network [38].

Cultivar	Maturity Classes (I_{AD})		
	Immature [1]	Harvest-Ready [2]	Mature [3]
'August Flame'	$I_{AD} > 1.3$	$1.3 \leq I_{AD} \leq 0.7$	$I_{AD} < 0.7$
'O'Henry'	$I_{AD} > 1.2$	$1.2 \leq I_{AD} \leq 0.7$	$I_{AD} < 0.7$
'Redhaven'	$I_{AD} > 1.6$	$1.6 \leq I_{AD} \leq 0.6$	$I_{AD} < 0.6$
'September Sun'	$I_{AD} > 1.2$	$1.2 \leq I_{AD} \leq 0.8$	$I_{AD} < 0.8$

[1] undetectable ethylene emission, [2] onset of ethylene emission and [3] ethylene emission peak.

2.4. Data Analysis and Modelling

2.4.1. Maturity Statistics

FD, FW, FF, SSC, I_{AD}, skin and flesh colour attributes (i.e., L*, a*, b*, C* and H°) were compared among the four cultivars using one-way analysis of variance (ANOVA), followed by the Games–Howell post hoc test for comparison of groups with unequal variances.

The correlations between pairs of the FD, FW, FF, SSC, I_{AD}, skin and flesh colour attributes was tested using Pearson's correlation coefficients (r) and presented in a correlation heatmap. From this point onwards, a* and H° were the only colour attributes considered for further data analyses, as b*, C* and L* are less important for yellow peach fruit characteristics.

Linear regression models of flesh a* and H° against I_{AD} were calculated to verify the validity of these two flesh parameters for maturity assessment. In addition, these models were used to estimate a* and H° maturity thresholds corresponding to I_{AD} maturity classes in the four peach cultivars (Table 1) in order to provide a clear range of a* and H° values for harvest-ready fruit, as these two parameters are not established in the stone fruit industry. ANOVA, post hoc tests and Pearson's correlations were carried out using R (v. 4.0.2) [39] and the "Userfriendliscience" package [40]. Graphs were generated using SigmaPlot 12.5 (Systat software Inc., Chicago, IL, USA).

2.4.2. Prediction Modelling

To achieve maturity prediction this work focused on analysing the fluorescence spectra using two different machine learning approaches. The first approach was used for the prediction of continuous variables such as FF, SSC, I_{AD}, skin and flesh a* and H°. The second approach was put in place to predict fruit maturity classes (as shown in Table 1) that represent simple categories that are widely understood by growers and industry stakeholders.

Prediction models for FF, SSC, I_{AD} and the most relevant colour attributes in both skin and flesh (a* and H°) were carried out using partial least square (PLS) regression with spectral band optimisation. Fluorescence spectra were not used for the prediction of FD and FW, as no relationship between fluorescence and fruit size was expected. All models were optimised using a k-fold cross-validation procedure using k = 10 splits. The basic idea was to select the most informative wavelength bands out of the entire spectral response. The selection was done via a random optimisation procedure called simulated annealing (SA) [41]. This algorithm begins with a randomly selected set of bands and, at each iteration, randomly changes a subset of these bands. A PLS regression model was developed at each iteration and the Akaike information criterion (AIC) [42] was used as the cost function. The aim of the algorithm was to decrease the cost functions, hence, to improve the model. The coefficients of determination for the cross-validation procedure (R^2_{CV}) and Lin's concordance correlation coefficients (r_c) [43] were calculated to assess models' robustness, with the latter being a reliable measure of the agreement between two variables. The r_c is particularly useful when comparing two measures of the same variable, such as when x (observed values) and y (predicted values) are in the same unit. The best models had the highest agreement between x and y and generated a r_c closer to 1 in a 0–1 range. The root mean square error in cross-validation ($RMSE_{CV}$) was also calculated to provide a measure of error in the same unit of the measured variable.

The SA optimisation procedure is a variant of a greedy optimisation strategy whereby the algorithm seeks to decrease the cost function, but not monotonically. SA allows for the occasional increase of the cost function to reduce the likelihood of the process getting stuck in a local minimum of the parameter space. In order for the SA process to be effective, the increase of the cost function must be occasional; in other words, it must happen with a small probability, governed by a hyper-parameter, which we chose to be equal to 0.1% of the current value of the AIC at each iteration.

The algorithm workflow was as follows:

1. Raw data were smoothed using convolution with a Gaussian kernel with $\sigma = 1.5$ units (corresponding to a physical sigma of approximately 3 nm at the centre of the spectrum).
2. Spectra were "re-binned" into 72 wavelength bands obtained by adding 4 contiguous wavelength bins.
3. The resulting spectra were then mean-centred and scaled to unit variance.
4. After these pre-processing steps the spectra were fed in the SA algorithm. The algorithm starts with a random draw of 20 bands (out of the total 72). At each iteration, a subset of two bands were randomly swapped, a PLS model was developed and cross-validated on the collection of bands selected by the SA. The optimum PLS model at each step was obtained by minimising the AIC. The algorithm was run for 5000 iterations.

The number of selected bands (20 in our case) was fixed for all models and initially chosen in accordance with the empirical optimal value of the latent variables selected by the cross-validation process. As such value was of the order of 10, we chose the number of bands to be double that, to provide some redundancy for the dimensionality reduction of the PLS algorithm. In other words, if the number of bands was to be further decreased, the number of latent variables tended to become similar to the number of bands, hence the model became over-constrained. R^2_{CV}, r_c, $RMSE_{CV}$, AIC and number of latent variables (LV) in the models were presented for the preferred PLS models.

Lastly, the selected 20 wavelength bands were also used to predict the fruit maturity classes (Table 1) in fruit of the four peach cultivars pooled together ($n = 800$) using a linear discriminant analysis (LDA). A three-class confusion matrix was used to visualise the performance of the model. The overall model's accuracy was based on the likelihood to predict the correct maturity class and expressed in F1-scores (0–1 range, where 1 is perfect prediction and 0 is no correct predictions) for each maturity class and for the whole model. The F1-score was calculated as the harmonic mean of precision and recall. Precision represents the number of true positive results divided by the sum of true and false positive results. Recall represents the number of true positive results divided by the sum of true positive results and false negative results. The model's error was calculated as $1 - F1$.

The algorithms were implemented in Python^{TM} 3.7.6 using the PLS regression and LDA routines available in the Scikit-learn package (v. 0.22.2) [44]. Sample scripts of the algorithms are freely available as a Project Jupyter Notebook [45].

3. Results

3.1. Fruit Maturity Indices

3.1.1. Comparisons between Cultivars

Maturity indices were found to be significantly different between the four yellow peach cultivars (Table 2). 'September Sun' had the largest FD and FW, whereas 'Redhaven' fruit were the smallest. Since 'Redhaven' fruit were harvested late, when already mature, their FF and I_{AD} were significantly lower than in other cultivars (Table 2). In 'August Flame', 'O'Henry' and 'September Sun', the median I_{AD} was within the respective harvest-ready I_{AD} range reported in Table 1. 'O'Henry' fruit had the highest sugars (median SSC of 14.8°Brix), whilst 'Redhaven' fruit were the least sweet (11.9°Brix).

Table 2. Fruit diameter (FD), fruit weight (FW), flesh firmness (FF), soluble solids concentration (SSC), I_{AD} and skin and flesh colour attributes (L*, a*, b*, C*, H°) in fruit of four yellow peach cultivars at harvest (n = 200). Medians and standard deviations (in brackets) are presented. Different letters represent significant differences ($p < 0.01$) between cultivars based on analysis of variance (ANOVA) followed by the Games–Howell post hoc test.

Maturity Index/Colour Attribute	Cultivar			
	'August Flame'	'O'Henry'	'Redhaven'	'September Sun'
FD (mm)	62.4 (5.5) c	64.9 (4.6) b	52.8 (5.4) d	70.2 (7.3) a
FW (g)	127.7 (30.4) c	146.7 (26.3) b	77.6 (24) d	186.3 (54.2) a
FF (kgf)	7.2 (1.8) a	6.2 (2.1) b	1.9 (1.5) c	6.1 (1.7) b
SSC (°Brix)	13.2 (2.3) b	14.8 (2.3) a	11.9 (1.3) c	13.4 (2.2) b
I_{AD}	1.1 (0.4) a	0.8 (0.5) b	0.1 (0.3) c	0.9 (0.4) b
Skin L*	48.0 (6.4) c	51.9 (7.1) b	56.2 (10.7) a	52.0 (8.3) b
Skin a*	27.5 (6.5) b	28.1 (6.4) b	30.6 (7.7) a	29.8 (6.7) ab
Skin b*	25.8 (8.3) b	25.9 (9.0) b	30.8 (11.6) a	29.9 (9.1) a
Skin C*	39.1 (8.5) c	40.2 (8.2) c	47.6 (9.3) a	43.5 (7.8) b
Skin H°	40.1 (9.7) b	39.4 (11.0) ab	42.4 (13.7) a	42.4 (11.4) ab
Flesh L*	78.5 (2.8) b	78.4 (3.1) bc	81.3 (6.5) a	77.8 (2.7) c
Flesh a*	4.9 (4.3) bc	7.4 (5.6) ab	6.7 (3.1) a	3.2 (3.8) c
Flesh b*	55.1 (3.5) c	55.9 (4.2) b	58.3 (4.2) a	54.8 (3.2) c
Flesh C*	55.6 (3.7) c	56.4 (4.5) b	58.7 (4.2) a	54.9 (3.4) c
Flesh H°	84.9 (4.3) ab	82.7 (5.5) bc	83.3 (2.9) c	86.5 (3.8) a

Skin colour was significantly different between the cultivars under study (Table 2). 'Redhaven' fruit had the significantly highest skin L* (i.e., lighter colouration), a* (i.e., redder than other cultivars in the green-red scale), b* (i.e., more yellow than other cultivars in the blue-yellow scale), C* (i.e., higher colour vividness than other cultivars) and H° (i.e., more orange-red than other cultivars in the red to yellow scale, i.e., 0 to 90°).

Flesh colour was also found to be significantly different between 'August Flame', 'O'Henry', 'Redhaven' and 'September Sun' (Table 2). Even in this case, the flesh of 'Redhaven' fruit had the highest L*, a*, b* and C*, but the lowest H° when compared to the other cultivars.

3.1.2. Correlations between Maturity Indices

Figure 1 highlights a colour-coded correlation strength between pairs of parameters, with darkest colours associated to strong correlations (blue for $r \geq 0.75$ and red for $r \leq -0.75$, $p < 0.001$) and white colour representing absent correlation ($-0.10 < r < 0.10$, $p > 0.05$). A strong correlation was observed between FW and FD in all the cultivars (Figure 1), as these two fruit size parameters are logically directly associated. In addition, strong correlation within skin colour attributes or within flesh colour attributes were not considered, as they are a consequence of collinearity among variables that are derived from their interdependency due to the fact that some of these variables are calculated from others (e.g., when a* increases, H° decreases, when C* increases, b* increases). SSC was not strongly correlated with other parameters ($r < 0.50$ or $>−0.50$), whereas FF and I_{AD} had an association with $r > 0.50$ in all the cultivars. FF and I_{AD} were in turn highly correlated ($r > 0.50$) with the flesh colour attributes a* and H°. Flesh a* was negatively correlated with I_{AD}, whereas flesh H° had a positive relationship.

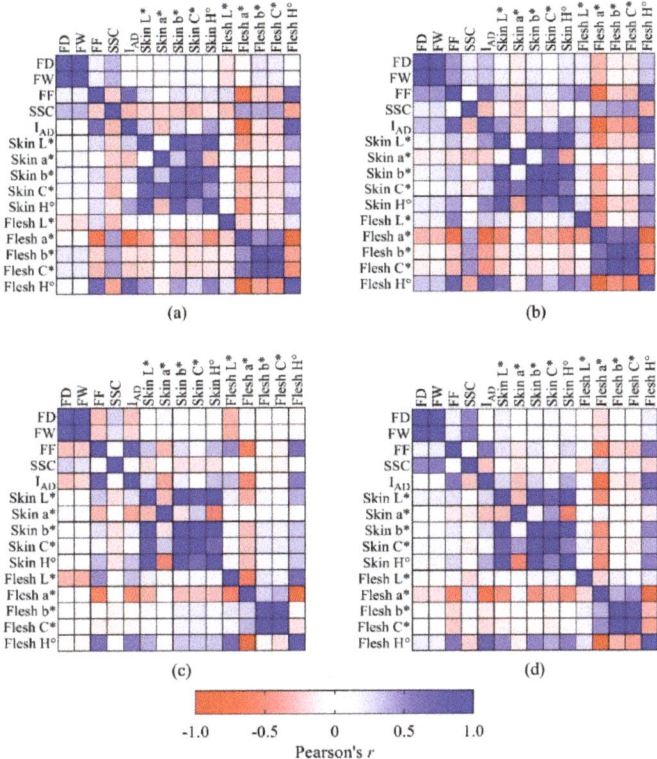

Figure 1. Pearson's correlation (r) heatmap for FD, FW, FF, SSC, I_{AD} and skin and flesh colour attributes (L*, a*, b*, C*, H°) in fruit of four yellow peach cultivars at harvest (n = 200). Correlation strength represented on a colour scale, where red represents high negative correlation, blue corresponds to high positive correlation and white is no correlation. (**a**) 'August Flame'; (**b**) 'O'Henry'; (**c**) 'Redhaven' and (**d**) 'September Sun'.

Although 'Redhaven' fruit were picked at a late maturity stage, correlations between parameters had in most cases similar strengths to those observed in 'August Flame', 'O'Henry' and 'September Sun' (Figure 1).

The scatter plots with linear fits in Figure 2 highlight the relationships of flesh a* and H° with I_{AD} in the four yellow peach cultivars. Table 3 shows linear regression models in which the I_{AD} maturity threshold values reported in Table 1 were set as x to predict a* and H° thresholds between maturity classes in the four cultivars under study. 'Redhaven's' linear regression model generated relatively low R^2 (<0.50) for both a* and H°, as expected from correlation heatmaps (Figure 1), whereas in the other cultivars the models' R^2 were higher than 0.50. The range of a* and H° values in harvest-ready fruit was relatively small in each of the four cultivars—from 2.6 ('September Sun') to 6.0 ('Redhaven') for a* and from 2.6 ('September Sun') to 5.8 ('Redhaven') for H°. Maturity thresholds appeared to be cultivar-specific, although 'August Flame' and 'O'Henry' fruit had almost identical a* and H° thresholds. The linear regression slopes were similar but inverse for a* and H°.

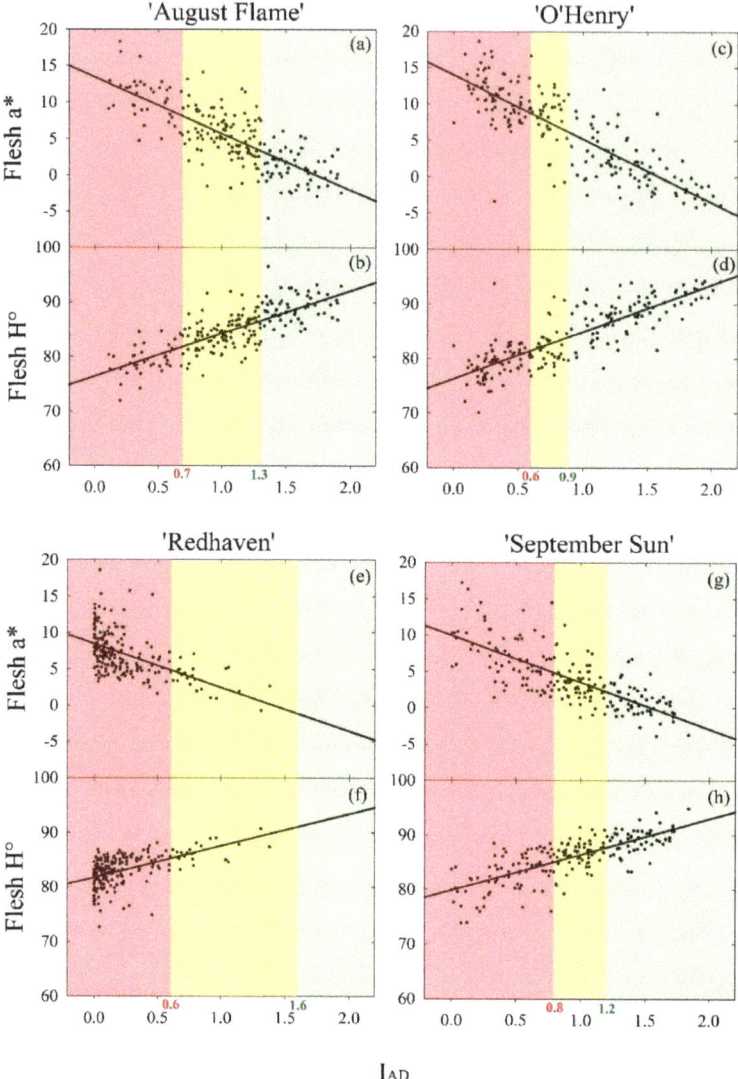

Figure 2. Scatter plots and linear regression fits of flesh a* and H° against the I_{AD} in fruit (n = 200) of four yellow peach cultivars at harvest (see Table 3 for regression coefficients). The three colours indicate I_{AD} maturity classes reported in Table 1: (green) immature—no ethylene emission; (yellow) harvest-ready—onset of ethylene emission and (red) mature—ethylene emission peak. (**a**) Flesh a* and (**b**) H° vs. I_{AD} in 'August Flame'; (**c**) Flesh a* and (**d**) H° vs. I_{AD} in 'O'Henry'; (**e**) Flesh a* and (**f**) H° vs. I_{AD} in 'Redhaven' and (**g**) Flesh a* and (**h**) H° vs. I_{AD} in 'September Sun'.

Table 3. Linear regression models and predicted thresholds (*y*) of flesh a* and H° for the maturity classes "harvest-ready" and "mature" based on the indices of absorbance difference (*x*) reported in Table 1. Data reported for fruit of four yellow peach cultivars at harvest (*n* = 200). Intercept, slope and coefficients of determination (R^2) reported for each linear regression model. Standard errors of intercept and slopes are in brackets.

Cultivar	Flesh Colour Attribute	Linear Regression Models			Predicted Maturity Thresholds	
		Intercept	Slope	R^2	Harvest-Ready	Mature
'August Flame'	a*	13.4 (0.5)	−7.8 (0.4)	0.60	3.4	8.0
	H°	76.4 (0.5)	7.8 (0.4)	0.62	86.6	81.9
'O'Henry'	a*	14.0 (0.4)	−8.8 (0.4)	0.70	3.5	7.8
	H°	76.3 (0.4)	8.6 (0.4)	0.71	86.7	82.3
'Redhaven'	a*	8.4 (0.2)	−6.0 (0.7)	0.28	−1.1	4.9
	H°	81.8 (0.2)	5.8 (0.6)	0.30	91.1	85.3
'September Sun'	a*	9.9 (0.4)	−6.4 (0.4)	0.55	2.2	4.8
	H°	79.9 (0.4)	6.5 (0.4)	0.57	87.7	85.1

3.2. Fluorescence Spectra and Maturity Prediction

3.2.1. Spectra Characteristics

To highlight the response of fluorescence emission under different maturity classes (based on the classification in Table 2) Figure 3 shows spectra of three individual 'August Flame' fruits. The immature and harvest-ready fruit emitted fluorescence peaks at ~680–690 nm, with the latter being sensibly lower than the former. However, the harvest-ready fruit showed higher emission between 450 and 650 nm compared to the immature fruit. The mature fruit had no clearly defined peaks, but instead showed a hump with maximum emission between 600 and 650 nm, a smaller but wider hump in the 480–550 nm range and a barely visible hump at 680–690 nm. Both the humps at 600–650 and 480–550 nm observed in the mature fruit were more pronounced than in the harvest-ready fruit.

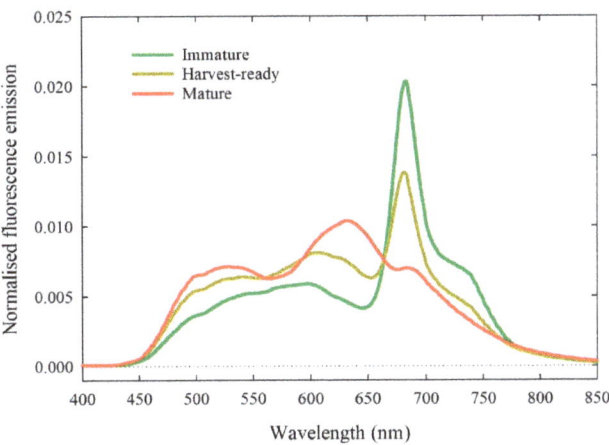

Figure 3. Normalised fluorescence emission spectra for three individual 'August Flame' fruit at different maturity classes. Immature: Index of absorbance difference (I_{AD}) > 1.3; harvest-ready: $1.3 \leq I_{AD} \leq 0.7$ and mature = I_{AD} < 0.7 (see Table 1).

Measured maturity indices corresponding to the fruit shown in Figure 3 are reported in Table 4. The immature fruit had the highest FF, SSC, I_{AD}, skin and flesh H°, and the lowest skin a* and flesh a*. This trend was expectedly inverted in mature fruit, in line with previous results (Figure 1).

Table 4. FF, SSC, I_{AD}, skin and flesh a* and H° in three individual 'August Flame' fruit shown in Figure 3.

Maturity Index/Colour Attribute	Immature	Harvest-Ready	Mature
FF [1] (kgf)	8.8	6.4	5.1
SSC [2] (°Brix)	12.7	12.2	12.4
I_{AD} [3]	1.5	1.0	0.4
Skin a*	25.7	32.3	35.2
Skin H°	53.7	37.3	31.6
Flesh a*	−2.5	5.0	8.9
Flesh H°	92.7	84.7	79.5

[1] Flesh firmness, [2] Soluble Solids Concentration, [3] Index of absorbance difference.

3.2.2. PLS Models

Once fluorescence emission spectra were processed with the PLS algorithm, predicted values of FF, SSC, I_{AD}, skin a* and H° and flesh a* and H° were obtained and compared to measured data. Figure 4 shows cross-validation predicted I_{AD} plotted against the observed I_{AD}. A visual assessment of the concordance between the two variables suggested that the PLS models provided a good estimation of I_{AD}, as model linear fits were relatively close to a $y = x$ reference line. Linear fits also show that the PLS models in the four cultivars slightly overestimated I_{AD} when its values were low—i.e., mature fruit—and underestimated them when they were high—i.e., immature fruit. However, overestimations at a minimum observed $I_{AD} = 0.0$ (+0.17, +0.15, +0.14 and 0.01 in 'O'Henry', 'August Flame', 'September Sun' and 'Redhaven', respectively) and underestimations at a maximum observed $I_{AD} = 2.2$ (−0.27, −0.20, −0.16 and −0.07 in 'O'Henry', 'September Sun', 'August Flame' and 'Redhaven', respectively) were considered very low. The model's linear fit in 'Redhaven' fruit was the closest to the $y = x$ line, although most of the points were found at low I_{AD}.

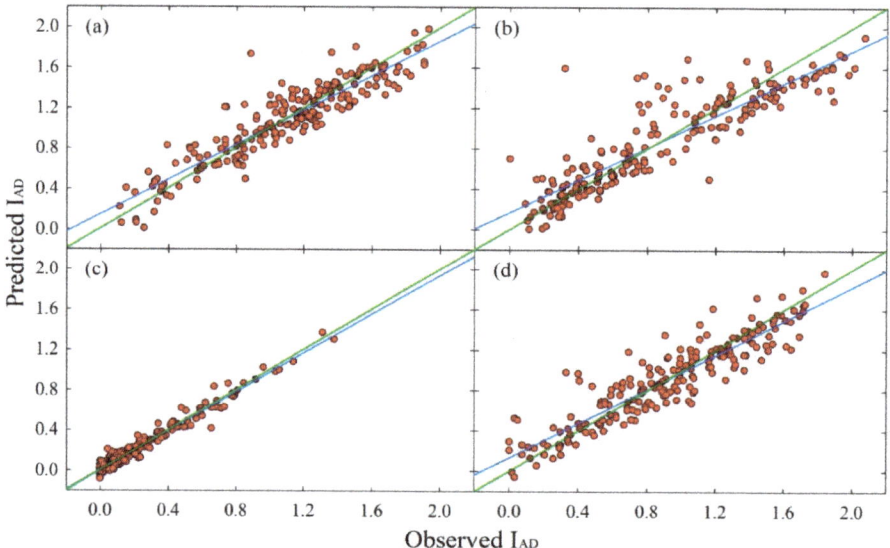

Figure 4. Scatter plots of cross-validation predicted I_{AD} against observed I_{AD} in yellow peach fruit (n = 200) using a partial least square regression (PLS) model: (**a**) 'August Flame'; (**b**) 'O'Henry'; (**c**) 'Redhaven' and (**d**) 'September Sun'. Blue lines represent partial least square regression model linear fits; green lines represent reference linear fits where predicted I_{AD} = observed I_{AD}. PLS statistics are reported in Table 5.

Table 5. Partial least square regression models for the prediction of FF, SSC, I_{AD}, skin and flesh a* and H° in four yellow peach cultivars ($n = 200$) using a custom-built fluorescence spectrometer.

Cultivar	Maturity Index	R^2_{CV} [1]	r_c [2]	AIC [3]	$RMSE_{CV}$ [4]	LV [5]
'August Flame'	FF [1]	0.69	0.82	184	1.48	12
	SSC [2]	0.72	0.84	301	1.9	12
	I_{AD} [3]	0.83	0.91	−655	0.18	10
	Skin a*	0.76	0.86	509	3.21	12
	Skin H°	0.77	0.87	657	4.65	11
	Flesh a*	0.76	0.87	337	2.09	12
	Flesh H°	0.80	0.89	305	1.93	12
'O'Henry'	FF	0.69	0.81	149	1.15	12
	SSC	0.75	0.86	196	1.55	11
	I_{AD}	0.80	0.89	−528	0.24	9
	Skin a*	0.75	0.85	467	3.21	7
	Skin H°	0.83	0.91	629	4.45	9
	Flesh a*	0.87	0.93	317	2.04	10
	Flesh H°	0.87	0.93	296	1.93	9
'Redhaven'	FF	0.78	0.87	−20	0.71	10
	SSC	0.58	0.74	220	1.34	7
	I_{AD}	0.96	0.98	−1126	0.05	9
	Skin a*	0.76	0.87	570	3.75	7
	Skin H°	0.88	0.94	668	4.78	10
	Flesh a*	0.46	0.62	375	2.26	9
	Flesh H°	0.49	0.65	340	2.07	10
'September Sun'	FF	0.50	0.67	280	1.95	6
	SSC	0.49	0.67	382	2.01	8
	I_{AD}	0.83	0.91	−651	0.18	8
	Skin a*	0.76	0.86	508	3.28	7
	Skin H°	0.79	0.88	692	5.21	10
	Flesh a*	0.73	0.84	312	2.01	10
	Flesh H°	0.74	0.85	303	1.97	7

[1] Coefficient of determination of the cross-validation, [2] Lin's concordance correlation coefficient, [3] Akaike information criterion, [4] root mean square error of the cross-validation and [5] number of latent variables.

The calculated R^2_{CV}, r_c and $RMSE_{CV}$ provided valuable information to better assess prediction power and error of the PLS models. A summary of these parameters is presented in Table 5 and highlight differences in models' robustness among cultivars. AIC values cannot be compared across cultivars and maturity indices, as they are only meant to be used to compare alternative models for the same dataset. The PLS model for FF and SSC predictions generated the overall lowest R^2_{CV} and r_c among maturity indices. The fluorescence emission spectra could only poorly predict SSC and FF in 'September Sun' fruit ($R^2_{CV} \leq 0.50$ and $r_c < 0.70$). I_{AD} was the only maturity index that was consistently and strongly predicted by the PLS models in all the cultivars ($R^2_{CV} \geq 0.80$ and $r_c \geq 0.89$), regardless of maturity advancement, with $RMSE_{CV} < 0.25$. R^2_{CV} and r_c were also consistently in the 0.75–0.88 and 0.85–0.94 ranges, respectively, in models for the prediction of skin a* and H°, with the latter being always the one with highest precision between the two-colour attributes. Conversely, the accuracy of flesh a* and H° prediction differed in the four cultivars, ranging from the most precise predictions in 'O'Henry' ($R^2_{CV} = 0.87$ and $r_c = 0.93$) to the poorest estimates in 'Redhaven' ($R^2_{CV} < 0.50$ and $r_c \leq 0.65$).

3.2.3. LDA Model

The LDA model for the pooled fruit dataset generated an overall $F1 = 0.85$; thus, the overall likelihood to estimate the correct maturity class was of approximately 85%, with an estimate error of 15%. The confusion matrix in Figure 5 shows that the model estimated 356, 185 and 135 true positives in mature, harvest-ready and immature fruit, when actual counts for the three classes were

387, 235 and 178, respectively. The single-class $F1$-scores were 0.93, 0.79 and 0.75 for mature, immature and harvest-ready fruit, respectively. The two opposite diagonal values in the black matrix cells show that only one mature fruit was classified as immature, and no immature fruit were classified as mature, providing additional strength to the model's results.

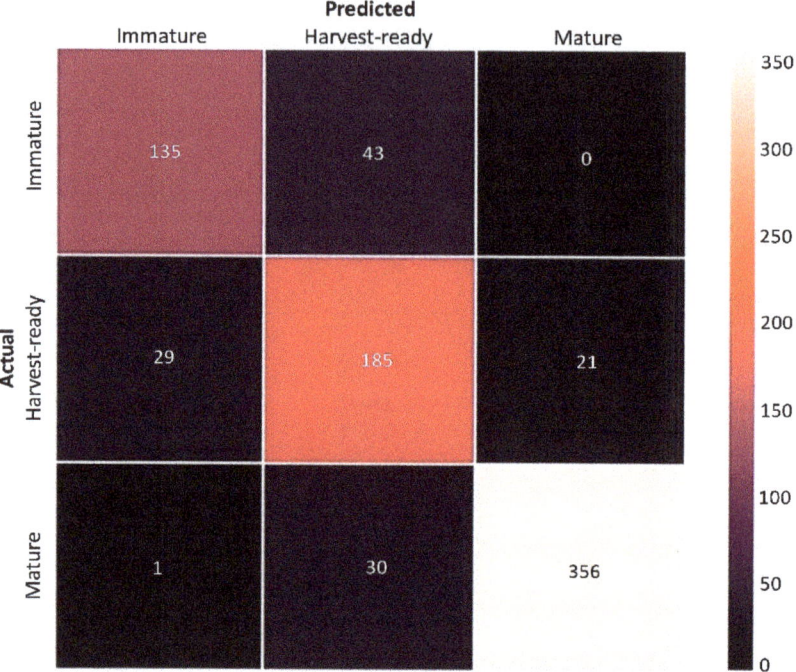

Figure 5. Confusion matrix of actual and predicted maturity classes generated after a linear discriminant analysis of the fluorescence spectra obtained from fruit of the four peach cultivars 'August Flame', 'O'Henry', 'Redhaven' and 'September Sun' pooled together ($n = 800$). Numbers in each cell represent fruit counts for a specific predicted-actual combination. Numbers in the diagonal top-left to bottom-right axis represent true positives of the prediction.

4. Discussion

The four cultivars selected for this study produced fruit with different characteristics, as highlighted in Table 2. The skin colour values reported for 'Redhaven' were partially influenced by advanced maturity due to the late sampling of fruit at harvest. However, skin colour was measured on the fruit cheek along the equatorial diameter, usually dominated by cover colour rather than ground colour, with the former being poorly influenced by maturity. Therefore, significantly different skin colouration among cultivars was most likely influenced by intrinsic genotypic characteristics. In flesh colour measurements, lower H° represented redder flesh, as also suggested by a* (Table 2), a typical consequence of fruit maturation. Therefore, flesh colouration differences in the four cultivars may be attributed to both genotypic characteristics and maturity stage.

The fact that SSC was not strongly correlated with other parameters (Figure 1) suggests that sugars are a relatively independent fruit quality parameter that should be used with caution when attempting to assess maturity in yellow peaches. The strong correlations between FF and I_{AD} indicate that there is a degree of interdependency between loss of firmness and chlorophyll degradation in yellow peaches. In addition, the strong correlation of chlorophyll degradation with the flesh colour attributes a* and H° was probably governed by the changes in flesh pigments, with an overall decrease of chlorophyll over the maturation period. An exception was highlighted in 'Redhaven', where the correlations between I_{AD} and flesh a* and H° were poorer than in other cultivars, probably influenced by the fact that when fruit are more mature, I_{AD} saturates towards its minimum value (i.e., 0.0) and fails to detect further physiological changes (e.g., cell wall and pectin modifications, internal darkening) that might instead be captured by variations in flesh colour attributes such as a* and H°. Our results support earlier work by Slaughter et al. [30], who observed a difference of approximately 5°H of flesh hue between immature and mature peaches. The linear models shown in Table 3 highlight that the flesh colour attributes a* and H° may be used as maturity indices, although thresholds between maturity classes are cultivar specific. The similar but inverse slope in their relationship with I_{AD} suggests that regression models based on one flesh colour attribute or the other are likely to have similar accuracy. This explains why literature shows the use of both a* and H° for fruit maturity assessments [30,46].

Typically, in immature fruit most of the fluorescence is emitted by chlorophyll pigments. Chlorophyll fluorescence has a primary peak at ~683 nm and a secondary peak at ~720 nm. The width of the peaks in stone fruit is usually comparable to their separation, so that the secondary peak appears as a shoulder between 700 and 750 nm [47] These features are clearly visible in immature and harvest-ready 'August Flame' fruit (Figure 3). In mature fruit, since chlorophyll had mostly degraded, its fluorescence emission is greatly diminished, and the measurable signal at ~683 nm is lower than the maximum emission emitted between 600 and 650 nm. The fluorescence emission observed at 600–650 nm is likely to be due to anthocyanins in the skin, as this range is compatible with fluorescence emission ranges of the 3-glucoside of malvidin (i.e., oenin) [48], an anthocyanin found in red grapes. Fluorescence emission at wavelength < 600 nm can be explained by a combination of flavonols and carotenoids. Flavonols such as quercetin, mostly found in yellow peach peels and scarcely present in flesh [49,50], emit green fluorescence with maxima near 520–530 nm [51]. Carotenoids have maximum fluorescence emission in the 450–550 nm range [52] and they are mainly present in the form of β-carotene and β-cryptoxanthin in yellow peach flesh [53]. On the one hand, the immature fruit showing a large chlorophyll peak at ~683 nm had, expectedly, the lowest skin and flesh a* and the highest skin and flesh H°, i.e., greener (Table 4). On the other hand, the mature fruit showed the highest emission in the flavonol and carotenoid spectral ranges (Figure 3), meaning that these pigments were the most likely candidates to induce high a* and low H° in both skin and flesh (i.e., redder) (Table 4). Changes of pigments concentrations in skin and flesh over fruit maturation are the basis for maturity prediction by fluorescence spectrometry.

PLS models highlighted that fluorescence emission can estimate FF and SSC in yellow peaches at harvest, although prediction is not very accurate (Table 5). I_{AD} and skin a* and H° were consistently predicted in all the cultivars with good accuracy ($R^2_{CV} \geq 0.75$ and $r_c \geq 0.85$). The prediction power for flesh a* and H° diverged among cultivars, with models for 'Redhaven'—the most mature fruit—showing low accuracy ($R^2_{CV} < 0.50$ and $r_c \leq 0.65$, Table 5). The high median values of skin L* and C* in 'Redhaven' fruit (Table 2) may have led to increased self-absorption of the fluorescence from the skin, hence, a decreased contribution of the pulp emission to the total measured fluorescence. This in turn, could have caused a reduction of the prediction ability of fluorescence spectrometry for flesh a* and H° in this cultivar.

The LDA model's F1-score was equal to 0.85, which represented a very good likelihood of predicting the correct maturity class. All the class predictions showed robust predictions ($F1 \geq 0.75$). The lowest F1 (0.75) was observed in the intermediate maturity class (harvest-ready) as in this case there were more chances to have a false positive on both sides of the class, i.e., neighbouring mature and immature classes. The highest accuracy obtained from the prediction of mature fruit ($F1 = 0.93$) was likely to be influenced by the larger sample size for this class ($n = 387$) compared to the other classes. The PLS models obtained for I_{AD} and the LDA models obtained for maturity classes satisfied the hypothesis of our study, as they generated maturity predictions with accuracies ≥ 0.85, in terms of r_c and F1-score, respectively.

The fluorescence spectrometer used in this study captures the steady-state fluorescence from the fruit under constant excitation light illumination. This system is fundamentally different from the pulse-amplitude modulation (PAM) fluorometers [54], which employ modulated excitation and fast detection to decouple fluorescence yield information from the measured fluorescence intensity. The sensitivity of a PAM fluorometer is derived from its time resolution and it is usually restricted to chlorophyll fluorescence. The fluorescence spectrometer described here does not produce fluorescence yield information, but it enables rapid measurements of a large spectral band without requiring specific sample preparation, most notably without requiring dark adaption of the fruit samples.

This characteristic is especially promising in view of field applications using hand-held devices, as the spectra can be captured without picking the fruit. The development of a wireless, battery-operated, handheld device for field estimation of fruit maturity is presently underway and early prototypes are being tested in field pilots of stone fruits (nectarine, white peach, plum, apricot) and pome fruits (apple, pear). While the current system gives good results in maturity estimation, the prediction of commercial maturity parameters such as SSC or FF is somewhat limited and very dependent on the cultivar. This suggests that a combination of fluorescence and NIR is required for improved performances. At the same time, a better modelling of the self-absorption properties of the fruit skin may help compensate from the signal reduction from the pulp, hence improve the model accuracy.

The ability to correctly predict maturity classes, though amenable to improvements, is suggestive of a post-harvest application of the fluorescence principle in graders. This application is currently being explored, in order to take advantage of prediction models that have the potential to be, to a large extent, independent from cultivar information within the same crop (e.g., yellow peach), hence better suited to a commercial operation with constantly changing cultivars being processed over a season.

5. Conclusions

In conclusion, the fluorescence spectrometer used in this study accurately predicted fruit maturity in yellow peaches. The machine learning prediction algorithms have the potential to be easily implemented in different configurations of the spectrometer (e.g., fluorescence alone, fluorescence + NIR, fluorescence + RGB and NIR in grading machines, portable or benchtop) for both in situ and post-harvest fruit maturity estimations. The sensor's applicability for fruit maturity prediction and for the assessment of fruit quality parameters and storage disorders in stone fruits and apples is currently under investigation at the Tatura SmartFarm.

Author Contributions: A.S., D.P. and M.G.O., conceptualization, methodology, investigation; A.S. and D.P., data collection & curation, visualization; A.S. and D.P., formal analysis, writing-original draft; A.S., D.P. and M.G.O., writing-review & editing; M.G.O., project management, reporting & administration, funding acquisition. All authors have read and agreed to the published version of the manuscript.

Funding: This study was financially supported by the following projects: (1) 'Horticulture Development Plan Task 1: Yield and quality relationships with light interception' funded by the Victorian Government's Agriculture Infrastructure and Jobs Fund; (2) 'SF17006: Summerfruit orchard—phase II' funded by Horticulture Innovation Australia Limited using Summerfruit levy and the Australian Government with co-investment from Agriculture Victoria; and (3) 'Deploying real-time sensors to meet Summerfruit export requirements for China' funded by Food Agility CRC Ltd., under the Commonwealth Government CRC Program. The CRC Program supports industry-led collaborations between industry, researchers and the community.

Acknowledgments: The technical support and assistance of Subhash Chandra, Dave Haberfield, Cameron O'Connell, Madita Lauer, Madeleine Peavey and Laura Phillips is gratefully acknowledged.

Conflicts of Interest: The authors declare no conflict of interest.

References

1. Crisosto, C.H. Stone fruit maturity indices: A descriptive review. *Postharvest News Inf.* **1994**, *5*, 65N–68N.
2. Kingston, C.M. Maturity Indices for Apple and Pear. In *Horticultural Reviews*; Janick, J., Ed.; John Wiley & Sons, Inc: Hoboken, NJ, USA, 1992; Volume 13, pp. 407–432. ISBN 978-0-471-57499-6.

3. Nicolaï, B.M.; Beullens, K.; Bobelyn, E.; Peirs, A.; Saeys, W.; Theron, K.I.; Lammertyn, J. Nondestructive measurement of fruit and vegetable quality by means of NIR spectroscopy: A review. *Postharvest Biol. Technol.* **2007**, *46*, 99–118. [CrossRef]
4. Betemps, D.L.; Fachinello, J.C.; Galarça, S.P.; Portela, N.M.; Remorini, D.; Massai, R.; Agati, G. Non-destructive evaluation of ripening and quality traits in apples using a multiparametric fluorescence sensor. *J. Sci. Food Agric.* **2012**, *92*, 1855–1864. [CrossRef] [PubMed]
5. Qin, J.; Chao, K.; Kim, M.S.; Lu, R.; Burks, T.F. Hyperspectral and multispectral imaging for evaluating food safety and quality. *J. Food Eng.* **2013**, *118*, 157–171. [CrossRef]
6. Bureau, S.; Cozzolino, D.; Clark, C.J. Contributions of Fourier-transform mid infrared (FT-MIR) spectroscopy to the study of fruit and vegetables: A review. *Postharvest Biol. Technol.* **2019**, *148*, 1–14. [CrossRef]
7. Ziosi, V.; Noferini, M.; Fiori, G.; Tadiello, A.; Trainotti, L.; Casadoro, G.; Costa, G. A new index based on vis spectroscopy to characterize the progression of ripening in peach fruit. *Postharvest Biol. Technol.* **2008**, *49*, 319–329. [CrossRef]
8. Infante, R. Harvest maturity indicators in the stone fruit industry. *Stewart Postharvest Rev.* **2012**, *8*, 1–6. [CrossRef]
9. Bonora, E.; Stefanelli, D.; Costa, G. Nectarine fruit ripening and quality assessed using the index of absorbance difference (IAD). *Int. J. Agron.* **2013**. [CrossRef]
10. McGlone, V.A.; Jordan, R.B.; Martinsen, P.J. Vis/NIR estimation at harvest of pre- and post-storage quality indices for "Royal Gala" apple. *Postharvest Biol. Technol.* **2002**. [CrossRef]
11. Fan, G.; Zha, J.; Du, R.; Gao, L. Determination of soluble solids and firmness of apples by Vis/NIR transmittance. *J. Food Eng.* **2009**. [CrossRef]
12. Palmer, J.W.; Harker, F.R.; Tustin, D.S.; Johnston, J. Fruit dry matter concentration: A new quality metric for apples. *J. Sci. Food Agric.* **2010**. [CrossRef] [PubMed]
13. Li, J.; Huang, W.; Zhao, C.; Zhang, B. A comparative study for the quantitative determination of soluble solids content, pH and firmness of pears by Vis/NIR spectroscopy. *J. Food Eng.* **2013**. [CrossRef]
14. Goke, A.; Serra, S.; Musacchi, S. Postharvest dry matter and soluble solids content prediction in d'anjou and bartlett pear using near-infrared spectroscopy. *HortScience* **2018**. [CrossRef]
15. Harker, F.R.; Carr, B.T.; Lenjo, M.; MacRae, E.A.; Wismer, W.V.; Marsh, K.B.; Williams, M.; White, A.; Lund, C.M.; Walker, S.B.; et al. Consumer liking for kiwifruit flavour: A meta-analysis of five studies on fruit quality. *Food Qual. Prefer.* **2009**. [CrossRef]
16. Bureau, S.; Ruiz, D.; Reich, M.; Gouble, B.; Bertrand, D.; Audergon, J.M.; Renard, C.M.G.C. Rapid and non-destructive analysis of apricot fruit quality using FT-near-infrared spectroscopy. *Food Chem.* **2009**. [CrossRef]
17. Escribano, S.; Biasi, W.V.; Lerud, R.; Slaughter, D.C.; Mitcham, E.J. Non-destructive prediction of soluble solids and dry matter content using NIR spectroscopy and its relationship with sensory quality in sweet cherries. *Postharvest Biol. Technol.* **2017**. [CrossRef]
18. Minas, I.S.; Blanco-Cipollone, F.; Sterle, D. Accurate non-destructive prediction of peach fruit internal quality and physiological maturity with a single scan using near infrared spectroscopy. *Food Chem.* **2021**. [CrossRef]
19. Scalisi, A.; O'Connell, M.G. Application of Visible/NIR spectroscopy for the estimation of soluble solids, dry matter and flesh firmness in stone fruits. *J. Sci. Food Agric.* **2020**. [CrossRef]
20. Chen, P.; McCarthy, M.J.; Kauten, R. NMR for internal quality evaluation of fruits and vegetables. *Trans. Am. Soc. Agric. Eng.* **1989**. [CrossRef]
21. Fuleki, T.; Cook, F.I. Relationship of Maturity as Indicated by Flesh Color to Quality of Canned Clingstone Peaches. *Can. Inst. Food Sci. Technol. J.* **1976**. [CrossRef]
22. Kader, A.A.; Heintz, C.N.; Chordas, A. Postharvest quality of fresh and canned peaches as influenced by genotypes and maturity at harvest. *J. Am. Soc. Hortic. Sci.* **1982**, *107*, 947–951.
23. Delwiche, M.J.; Baumgardner, R.A. Ground color as a peach maturity index. *J. Am. Soc. Hortic. Sci.* **1985**, *110*, 53–57.
24. Tourjee, K.R.; Barrett, D.M.; Romero, M.V.; Gradziel, T.M. Measuring flesh color variability among processing clingstone peach genotypes differing in carotenoid composition. *J. Am. Soc. Hortic. Sci.* **1998**, *123*, 433–437. [CrossRef]
25. Zhang, P.; Wei, Y.; Xu, F.; Wang, H.; Chen, M.; Shao, X. Changes in the chlorophyll absorbance index (IAD) are related to peach fruit maturity. *N. Z. J. Crop Hortic. Sci.* **2020**, *48*, 34–46. [CrossRef]

26. Greer, D.H. Non-destructive chlorophyll fluorescence and colour measurements of 'braeburn' and 'royal gala' apple (malus domestica) fruit development throughout the growing season. *N. Z. J. Crop Hortic. Sci.* **2005**, *33*, 413–421. [CrossRef]
27. ISO 11664-4:2008(en), Colorimetry—Part 4: CIE 1976 L*a*b* Colour Space. Available online: https://www.iso.org/obp/ui/#iso:std:iso:11664:-4:ed-1:v1:en (accessed on 24 August 2020).
28. McGuire, R.G. Reporting of Objective Color Measurements. *HortScience* **1992**, *27*, 1254–1255. [CrossRef]
29. Peavey, M.; Goodwin, I.; McClymont, L.; Chandra, S. Effect of shading on red colour and fruit quality in blush pears "ANP-0118" and "ANP-0131". *Plants* **2020**, *9*, 206. [CrossRef] [PubMed]
30. Slaughter, D.C.; Crisosto, C.H.; Hasey, J.K.; Thompson, J.F. Comparison of instrumental and manual inspection of clingston peaches. *Appl. Eng. Agric.* **2006**, *22*, 883–889. [CrossRef]
31. Ruiz, D.; Egea, J.; Tomás-Barberán, F.A.; Gil, M.I. Carotenoids from new apricot (*Prunus armeniaca* L.) varieties and their relationship with flesh and skin color. *J. Agric. Food Chem.* **2005**, *53*, 6368–6374. [CrossRef]
32. Slaughter, D.C.; Crisosto, C.H.; Tiwari, G. Nondestructive determination of flesh color in clingstone peaches. *J. Food Eng.* **2013**, *116*, 920–925. [CrossRef]
33. Belefant-Miller, H.; Miller, G.H.; Rutger, J.N. Nondestructive Measurement of Carotenoids in Plant Tissues by Fluorescence Quenching. *Crop Sci.* **2005**, *45*, 1786–1789. [CrossRef]
34. Merzlyak, M.N.; Melø, T.B.; Naqvi, K.R. Effect of anthocyanins, carotenoids, and flavonols on chlorophyll fluorescence excitation spectra in apple fruit: Signature analysis, assessment, modelling, and relevance to photoprotection. *J. Exp. Bot.* **2008**, *59*, 349–359. [CrossRef]
35. Scalisi, A.; O'Connell, M.G.; Lo Bianco, R. Field non-destructive determination of nectarine quality under deficit irrigation. *Acta Hortic.* **2021**, in press.
36. Merzlyak, M.N.; Solovchenko, A.E.; Gitelson, A.A. Reflectance spectral features and non-destructive estimation of chlorophyll, carotenoid and anthocyanin content in apple fruit. *Postharvest Biol. Technol.* **2003**, *27*, 197–211. [CrossRef]
37. Solovchenko, A.; Chivkunova, O.; Gitelson, A.; Merzlyak, M. Non-Destructive Estimation Pigment Content, Ripening, Quality and Damage in Apple Fruit with Spectral Reflectance in the Visible Range. *Fresh Prod.* **2010**, *4*, 91–102.
38. Frisina, C.; Bonora, E.; Ceccarelli, A.; Stefanelli, D. DA Meter IAD Maturity Classes: Database—HIN. Available online: http://www.hin.com.au/networks/profitable-stonefruit-research/stonefruit-maturity-and-fruit-quality/da-meter-iad-maturity-classes-database (accessed on 24 August 2020).
39. R Core Team. *R: A Language and Environment for Statistical Computing*; R Foundation for Statistical Computing: Vienna, Austria, 2018; Available online: https://www.R-project.org/ (accessed on 7 September 2020).
40. Peters, G. *Userfriendlyscience: Quantitative Analysis Made Accessible*; R Package Version 0.7.2; 2018; Available online: https://userfriendlyscience.com (accessed on 7 September 2020).
41. van Laarhoven, P.J.M.; Aarts, E.H.L. Simulated annealing. In *Simulated Annealing: Theory and Applications*; Springer: Dordrecht, The Netherlands, 1987; pp. 7–15. ISBN 978-94-015-7744-1.
42. Akaike, H. A New Look at the Statistical Model Identification. *IEEE Trans. Automat. Contr.* **1974**, *19*, 716–723. [CrossRef]
43. Lin, L.I.-K. A Concordance Correlation Coefficient to Evaluate Reproducibility. *Biometrics* **1989**, *45*, 255–268. [CrossRef]
44. Pedregosa, F.; Varoquaux, G.; Gramfort, A.; Michel, V.; Thirion, B.; Grisel, O.; Blondel, M.; Prettenhofer, P.; Weiss, R.; Dubourg, V.; et al. Scikit-learn: Machine learning in Python. *J. Mach. Learn. Res.* **2011**, *12*, 2825–2830.
45. Pelliccia, D. Wavelength Band Selection with Simulated Annealing. Available online: https://github.com/nevernervous78/nirpyresearch/blob/master/snippets/Wavelengthbandselectionwithsimulatedannealing.ipynb (accessed on 7 September 2020).
46. do Nascimento Nunes, M.C. *Color Atlas of Postharvest Quality of Fruits and Vegetables*; Wiley-Blackwell: Ames, IA, USA, 2008.
47. Pedrós, R.; Moya, I.; Goulas, Y.; Jacquemoud, S. Chlorophyll fluorescence emission spectrum inside a leaf. *Photochem. Photobiol. Sci.* **2008**, *7*, 498–502. [CrossRef]
48. Agati, G.; Matteini, P.; Oliveira, J.; De Freitas, V.; Mateus, N. Fluorescence approach for measuring anthocyanins and derived pigments in red wine. *J. Agric. Food Chem.* **2013**, *61*, 10156–10162. [CrossRef] [PubMed]

49. Tomás-Barberán, F.A.; Gil, M.I.; Cremin, P.; Waterhouse, A.L.; Hess-Pierce, B.; Kader, A.A. HPLC—DAD—ESIMS analysis of phenolic compounds in nectarines, peaches, and plums. *J. Agric. Food Chem.* **2001**, *49*, 4748–4760. [CrossRef] [PubMed]
50. Scordino, M.; Sabatino, L.; Muratore, A.; Belligno, A.; Gagliano, G. Phenolic Characterization of Sicilian Yellow Flesh Peach (*Prunus persica* L.) Cultivars at Different Ripening Stages. *J. Food Qual.* **2012**, *35*, 255–262. [CrossRef]
51. Lang, M.; Stober, F.; Lichtenthaler, H.K. Fluorescence emission spectra of plant leaves and plant constituents. *Radiat. Environ. Biophys.* **1991**, *30*, 333–347. [CrossRef]
52. Zaghdoudi, K.; Ngomo, O.; Vanderesse, R.; Arnoux, P.; Myrzakhmetov, B.; Frochot, C.; Guiavarc'h, Y. Extraction, Identification and Photo-Physical Characterization of Persimmon (*Diospyros kaki* L.) Carotenoids. *Foods* **2017**, *6*, 4. [CrossRef]
53. Gil, M.I.; Tomás-Barberán, F.A.; Hess-Pierce, B.; Kader, A.A. Antioxidant capacities, phenolic compounds, carotenoids, and vitamin C contents of nectarine, peach, and plum cultivars from California. *J. Agric. Food Chem.* **2002**, *50*, 4976–4982. [CrossRef]
54. Schreiber, U. Pulse-Amplitude-Modulation (PAM) Fluorometry and Saturation Pulse Method: An Overview. In *Chlorophyll a Fluorescence. Advances in Photosynthesis and Respiration*; Papageorgiou, G.C., Govindjee, Eds.; Springer: Dordrecht, The Netherlands, 2004; Volume 19, pp. 279–319.

Publisher's Note: MDPI stays neutral with regard to jurisdictional claims in published maps and institutional affiliations.

© 2020 by the authors. Licensee MDPI, Basel, Switzerland. This article is an open access article distributed under the terms and conditions of the Creative Commons Attribution (CC BY) license (http://creativecommons.org/licenses/by/4.0/).

Article

Validation of Real-Time Kinematic (RTK) Devices on Sheep to Detect Grazing Movement Leaders and Social Networks in Merino Ewes

Hamideh Keshavarzi [1,*], Caroline Lee [1], Mark Johnson [2], David Abbott [2], Wei Ni [2] and Dana L. M. Campbell [1]

[1] Agriculture and Food, Commonwealth Scientific and Industrial Research Organisation (CSIRO), Armidale, NSW 2350, Australia; Caroline.Lee@csiro.au (C.L.); Dana.Campbell@csiro.au (D.L.M.C.)
[2] Data61, Commonwealth Scientific and Industrial Research Organisation (CSIRO), Marsfield, NSW 2122, Australia; Mark.Johnson@data61.csiro.au (M.J.); David.A.Abbott@data61.csiro.au (D.A.); Wei.Ni@data61.csiro.au (W.N.)
* Correspondence: Hamideh.Keshavarzi@csiro.au

Academic Editor: Spyridon Kintzios
Received: 17 December 2020; Accepted: 27 January 2021; Published: 30 January 2021

Abstract: Understanding social behaviour in livestock groups requires accurate geo-spatial localisation data over time which is difficult to obtain in the field. Automated on-animal devices may provide a solution. This study introduced an Real-Time-Kinematic Global Navigation Satellite System (RTK-GNSS) localisation device (RTK rover) based on an RTK module manufactured by the company u-blox (Thalwil, Switzerland) that was assembled in a box and harnessed to sheep backs. Testing with 7 sheep across 4 days confirmed RTK rover tracking of sheep movement continuously with accuracy of approximately 20 cm. Individual sheep geo-spatial data were used to observe the sheep that first moved during a grazing period (movement leaders) in the one-hectare test paddock as well as construct social networks. Analysis of the optimum location update rate, with a threshold distance of 20 cm or 30 cm, showed that location sampling at a rate of 1 sample per second for 1 min followed by no samples for 4 min or 9 min, detected social networks as accurately as continuous location measurements at 1 sample every 5 s. The RTK rover acquired precise data on social networks in one sheep flock in an outdoor field environment with sampling strategies identified to extend battery life.

Keywords: RTK u-blox; accuracy; sampling rate; social networks; leadership

1. Introduction

Sheep are social animals that live in groups and rely on social mechanisms to enhance their survival. They show sophisticated social behaviour with the ability to recognise faces of individual flock mates over extended periods of time [1], recognise conspecific faces that exhibit fear/stress [2], show differences in dominance relationships [3] and individuals will utilise herd protection while under a predator threat [4]. Measuring the social relationships and/or network behaviour of sheep can provide an understanding of leader influences in daily movement patterns [5], how social bonds may affect grazing patterns [6–8], how temperament, age, weather, and management practices affect social relationships [9,10], and how differences in gregariousness can impact group behavioural synchronisation [11]. Additionally, changes in social patterns may be used as an indication of some forms of distress in the individuals and/or flock [12]. However, traditional methods of data collection, such as live observations or decoding video recordings, can be labour-intensive, logistically challenging, and subjective which may limit the understanding of relationships that could be present. Setting up video camera systems in commercial farms, for example, can include challenges such as stocking density and variable lighting and background [13]. Live observations are limited by personnel availability, restricted to certain time windows (typically daytime), and human presence may affect normal animal behaviour.

With the development of on-animal sensors and technologies such as GPS devices [14,15], proximity loggers [9,16], or ultra-wideband positional loggers [17,18] there is the potential to deploy these devices on livestock individuals within groups to enable more accurate monitoring of position and/or social relationships. The data from these sensors can provide new insights into how individuals in a group interact and/or influence each other, including affiliative and/or agonistic relationships between group members [19,20], network structure [19], and resource-use patterns [21]. Commercially, sensor technologies that allow quantification of social interactions may, for example, enable understanding of how animals learn new technologies such as virtual fencing [22], detect male-female interactions to determine oestrus [23] and mating [24], allow maternal pedigree detection through ewe-lamb contact [25], or could monitor grazing behaviour [26]. Thus, the potential for the application of sensors to detect social interactions, leaders and networks is fast developing and can provide new insights into the social behaviour of livestock animals. Specifically, for sheep, there are continually increasing numbers of studies validating on-animal sensor applications for measures such as behaviour, health, and environmental management [27].

When deploying devices onto animals, the spatial precision of the sensors is important for monitoring the animal's behaviour or grouping the animals into sub-groups. In social network analysis, for example, the social structures identified by statistical processes are influenced by the way that data are collected among individuals [28]. In proximity-based social networks (PBSNs), the network is created based on close proximity between individuals which relies on spatial location data to create the network [29]. Therefore, more frequent and accurate data collection to capture all possible social interactions will result in more precise sub-groupings and allow detection of the social network structure within the group. In addition, more precise data may enable researchers to quantify interactions in situations where animals are all in close physical contact [20]. While GPS tracking has been used to quantify social relationships between livestock animals [30–32], GPS typically has a high spatial precision error; one study showed an overestimation of 15.2% or 1.5 km for daily cattle travels without any data filtering [33], another study showed a contact distance error of 9.5 m with prototype proximity-logging GPS collars on bighorn sheep [34]. These errors may limit detection accuracy of social associations. Obtaining positional data from multiple satellite systems will increase positional accuracy [35] which may then improve the accuracy of social network analyses, but research is currently limited. Haddadi et al. conducted a study [36] to measure social networks in sheep using data loggers which were custom designed GPS devices with the ability to record the phase shift of GPS signals at a rate of 1 Hertz, with an estimated accuracy of 20 cm achieved by applying Real-Time-Kinematic (RTK) corrections to the GPS signal during post processing. They found the GPS device effective at accurately characterising the network structure in a mixing experiment with Merino sheep. Normal Global Navigation Satellite System (GNSS) operation, which utilises multiple satellite constellations, has location errors of a few meters in clear locations but tens of meters under challenging environments. With RTK GNSS which is a normal GNSS operation improved for carrier phase tracking of the satellite signals and differential correction, the errors are reduced to the order of tens of centimeters in open environments with good satellite visibility. The accuracy achieved will depend on the exact technology that is used [37]. Under challenging conditions, the RTK correction signal may be blocked to varying degrees by buildings, dense tree coverage, or other animals and may require additional data collection such as IMU (inertial measurement unit) to improve geo-spatial data accuracy. An RTK device may enable highly accurate geo-spatial tracking of animals for precise detection of social networks.

Further practical considerations for long-term deployment of sensor devices include the issue of power consumption. One way to overcome this problem and extend the device's battery life is to decrease the sampling rate, but this would likely result in missing data which could impact the accuracy of the social network analysis [38]. Additionally, a misrepresentation of network properties in a simulated animal social network as a result of using incomplete information (edge sample size) has been reported by Perreault, 2010 [39]. Thus, the appropriate rate of sampling and its impact on sheep social network analysis requires further investigation.

This study validated a novel RTK device that attains accuracy that cannot be achieved by GPS alone, to automatically track individual animals kept in groups. Specifically, the current study aimed to:

1. Test the accuracy of the RTK devices in terms of consistency and error points first in the laboratory and then later in the field using a group of sheep.
2. Validate the devices for identifying leaders based on sheep movement data during a grazing period.
3. Validate that the generated GNSS positional data could be used to detect social networks in sheep.
4. Determine the optimal sampling rates to extend the battery life but still identify sheep social networks.

Validation of these on-animal sensor devices could enable more accurate automated data collection to understand livestock social behaviour in future research.

2. Materials and Methods

2.1. Design of the RTK-GNSS Device

The RTK-GNSS localisation device used in this study was based on an RTK module (model: C94-M8P) manufactured by u-blox (Thalwil, Switzerland). This RTK module can function as either an RTK base station, or as an RTK mobile rover. The C94-M8P module comprises a 72-channel GNSS receiver and an unlicensed band 433 MHz radio transceiver.

For this study, each sheep was fitted with an RTK rover. The rovers augmented the RTK module with a power pack (5 V 10,000 mAh), a micro SD card (32 GB), and an ARM single board computer (Figure 1a). The ARM processor was a SparkFun 9DoF Razor IMU MO with a SAMD21 microprocessor, an MPU-9250 nine degree of freedom inertial sensor and an SD card socket. The rover module was completed by an external GNSS active antenna, and a communications antenna to receive the correction messages from the base station. In the field testing with animals (see Section 2.3), the rover electronics modules were mounted in a box of dimensions 145 mm L × 105 mm W × 68 mm H, and a total weight of 607 gm with the communications antenna raised 127 mm above the top of the box (Figure 1b).

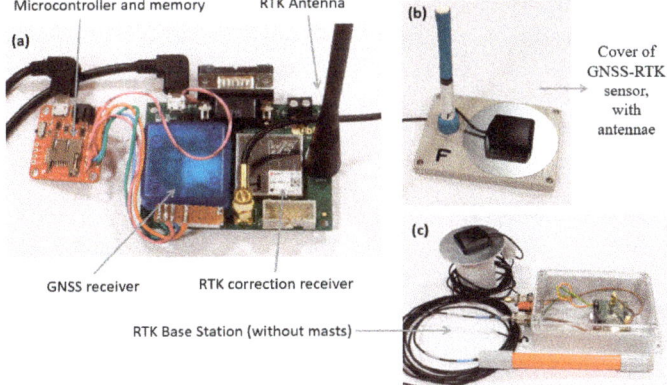

Figure 1. The system hardware components consisting of: (**a**) rover electronics showing the data logger, and Real-Time-Kinematic (RTK) module, (**b**) the rover packaging showing the Global Navigation Satellite System (GNSS)) and RTK correction antennae, and (**c**) the base station with the electronics and antennae (without their masts).

A single static RTK base station was placed in the middle of the test paddock. The base station required an RTK module, a GNSS antenna, a communications antenna (Figure 1c), and a power supply consisting of a 12 V battery and a solar battery charger.

The GNSS system was configured to use two civilian band satellite constellations, GPS, and GLONASS. The RTK base station received continuous GNSS signals (the purple lines in Figure 2) that were compared with the GNSS signals expected at the base station given the known location of the base station. The delay error in the GNSS signal between each visible GNSS satellite and the base station was calculated and broadcast on the 433 MHz radio as a correction message (red arrows in Figure 2). Each rover received GNSS signals and the correction data applied the delay correction to the GNSS signals and calculated the rover position. Each rover recorded to the SD card the GNSS time, location, and the quality of the RTK correction at a one Hertz rate, and inertial measurements at a 50 Hertz rate.

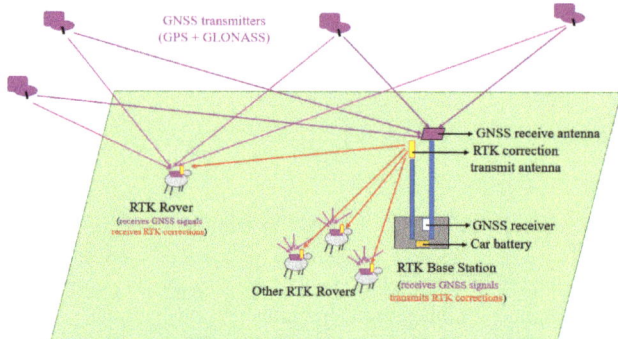

Figure 2. System operational diagram showing GNSS satellites, the RTK base station and RTK rovers harnessed to sheep. The purple lines are the continuous GNSS signals sent to the base station, and red arrows are the computed and transmitted correction messages from base station to RTK rovers.

2.2. Preliminary Laboratory Testing of System Performance

Before the field implementation on sheep, two rounds of preliminary testing were carried out in Marsfield, NSW, Australia to ensure the performance of the system. The first test was to determine the operational range of the RTK GNSS system. The base station and rover were mounted on a portable table at one end of the street where the test was done, and in a car, respectively. The car was driven away from and back toward the base station to test the communication range. The communication range was recorded with both base station and rover antenna vertical as well as with the base station antenna mounted with a 40-degree tilt towards the car, and a 40-degree tilt orthogonal to the car (Figure 3a). This tilt simulated the effect of antenna rotation due to the movements and positions of the sheep. The line of sight (LOS) range of the 433 MHz radio signal was limited to less than three hundred metres by the topology of the road and surrounding foliage.

The second test observed differential accuracy of multiple RTK rovers in a dynamic environment, which was relevant to the application on sheep. Six RTK rovers were attached at one metre separation on a two by three grid frame being carried between two experimenters. The plot in Figure 3b shows the base station (red/yellow star), and the six location tracks of the rovers. The experimenters started walking in the mid-left (labelled GNSS lock), heading SE. Because the rovers were fixed to the frame, any relative location errors were due to the system and the environment. The 35-min experiment included: (1) static periods to look at system noise performance, (2) walking periods with a line of sight between the base station, the rovers, and the GNSS constellations, and (3) periods of blockage between the rovers and the sky and non-line of sight (NLOS) between the rovers and the base station. The frame was placed on stands for 12 min in the centre south, then walked NE and placed on stands for another 10 min, then walked SW-SE-NE and in a loop into the trees in the right-hand side of the test field. The frame was walked out of the trees and placed back on the stands for 2 min, before walking NW towards the building in the centre of Figure 3b. The remaining walk followed a concrete path (NW-SW-NE-loop-NW) and finally along a road behind some trees to test range and signal blockage NE-SW.

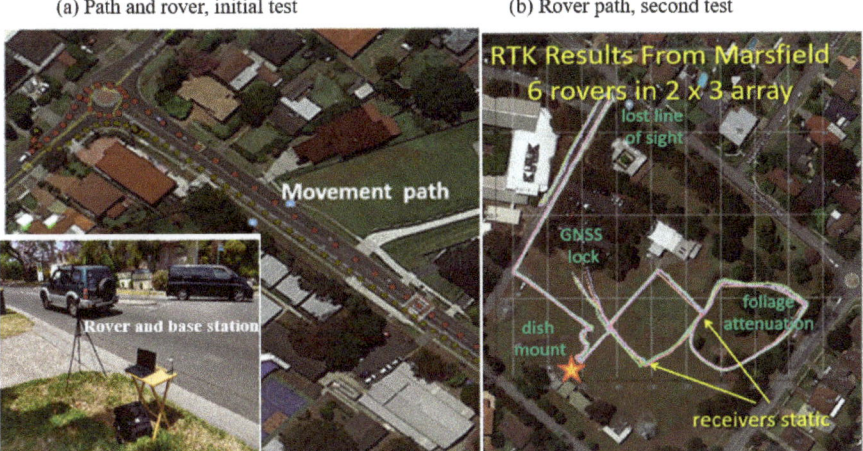

Figure 3. The areas where the accuracy performance of the device was first tested including (**a**) the initial test with the rover driven in a car (movement track indicated by the red dotted line), and (**b**) the second test with 6 rovers carried by experimenters (movement tracks indicated by coloured lines) where 'GNSS lock' indicates the starting position and the star indicates the position of the base station. Note: The multiple lines indicating 6 tracks can be seen during the initial GNSS lock up as well as in the "foliage attenuation" area, but the accuracy of the results negates clear visual separation of the 6 tracks during optimal operation.

2.3. Field Implementation on Sheep

2.3.1. Ethical Statement

The experiment with animals was approved by the CSIRO FD McMaster Laboratory Chiswick Animal Ethics Committee (ARA 19-27).

2.3.2. Animals and Experimental Protocol

Seven 1-year old Merino ewes (average body weight of 36.9 ± 5.9 kg) were used in this experiment. The animals were selected randomly from a research flock located at the Commonwealth Scientific and Industrial Research Organisation (CSIRO) Chiswick Research Station (Armidale, NSW, Australia) and had no previous experience with wearing GPS-devices. Two days before data collection commenced, dog harnesses (Comfy Harness, size 8, 84–120 cm, Company of Animals, Surrey, UK) were placed on the sheep to habituate them to the equipment. On the day of the study, an RTK rover was fitted to the dog harness and secured using plastic netting and cable ties. The harness was then fitted onto the backs of the sheep (see Figure 4a). For ease of checking, each sheep was numbered with coloured sheep wool marker (Heiniger Shearing Supplies, Briba Lake, WA, Australia) that matched different coloured antennas on the RTK rovers. Animals were placed into a paddock approximately 100 m × 70 m in size. The sheep had been kept in the same paddock as a group for four weeks prior to the study so were familiar with each other. The paddock was estimated to have approximately 2500 kg DM/ha of pasture available, and water available at the NE and SW corners of the paddock. The study was conducted across four consecutive days from 11 to 14 February 2020 (summer season). During the study period, two days (second and third day) experienced some intermittent rain and the weather was cloudy on the other two days. The mean minimum, overall, and maximum temperatures across 24 h periods over the test days were: mean ± SEM min: 21.25 ± 0.12 °C, avg: 24.62 ± 0.49 °C, max: 28.00 ± 0.54 °C based on weather data collected directly at the Chiswick site.

Figure 4. RTK rovers fitted on the back of sheep via harnesses (**a**), and fixed base station (**b**) in the middle of the paddock. The two masts (**b**) were for the GNSS amplified antenna, and RTK correction transmitter.

On each of the four study days, the devices were attached to the animals in the morning and then removed at the end of each day after 5 h of testing. The device continually recorded the GNSS location data throughout the daytime for 5 h per day with a sampling rate of one second. The GNSS receivers were removed from the sheep each evening. Sheep were kept in a small yard overnight with free access to water but not food to encourage grazing during the day. Sheep were checked twice daily during each day of testing to ensure the devices remained in position. On the last day, RTK rover D slipped to the side so all animals were brought into the yards at 10:30 a.m. to fix it before being placed back into the test paddock at 10:39 a.m. In addition, animals were video-recorded using a hand-held video recorder (Sony Handycam, HDR-XR260E, Sony Electronics Inc., Tokyo, Japan) twice per day for an hour each time; one hour in the morning soon after the animals were released into the paddock (8:00–9:00 a.m. except for the first day which was 11:20 a.m.–12:20 p.m. due to a late starting time) and one hour in the afternoon (2:00–3:00 p.m. for the first day and 1:00–2:00 p.m. for the remainder) for ground proofing of the data collected from the RTK Rovers. On the first day of the experiment, the reference base station was fixed in the middle of the paddock (Figure 4b) and the location determined from GNSS averaging. The range of the 433 MHz communication radio was measured to ensure that the signal covered the test paddock. To estimate the dynamic accuracy of the devices, a person walked around the paddock fence line holding one RTK rover.

2.4. Data Analyses

2.4.1. Device Accuracy and Reliability

Data from day one were incomplete and were not used for the calculation of social networks. RTK rover G was not used on day one after physical damage during transport. Data analysis showed that RTK rover C had temporary recording failures. Data from day four were not analysed for social networks as RTK rover F failed to record on this day due to operator failure (the device was not accurately started). Data from days two and three were complete and were used for calculations of social networks and sampling rates as described in subsequent sections. The analysis of the GNSS data recorded by the RTK rover was carried out using R packages [40]. All records commencing from when the animals were introduced into the paddock until they were removed remained in the analysis, including the GNSS error around the paddock boundaries to evaluate the GNSS accuracy. All available locational data for animals were plotted in the R statistical package per day (5 h a day).

The fence line coordinates were also plotted using ggplot [41] in R [40] based on the measurements obtained while walking around the fence line. To examine the device accuracy, position differences were calculated based on the difference between the instantaneous measurement from a GNSS rover moving along the paddock's fence line, compared with the fence line interpolated from a GNSS survey of the paddock corners. The calculated location errors were then plotted using ggplot [41] in R [40] with the normalised counts (i.e., count in each bin divided by total count so all values are <1.0) for all bins to make the comparison easier.

2.4.2. Identifying Leaders from Movement Patterns

Detection of leading sheep during movement around the paddock was based on the GNSS data showing animal movement during the grazing period. This period was within the first two hours of introducing animals into the paddock. The collected data during days 2 and 3 of the study were used as all sensors worked well for these two days as mentioned in Section 2.4.1. The first two hours of the day were selected as animals displayed their normal diurnal behaviour of movement (based on their graphical movement plots) and grazing following a period of overnight feed restriction (although visual confirmation of grazing was not applied for the entire 2-h period (see Section 2.4.4 for video records of behaviour for a portion of the grazing period). Each two-hour period within each study day was divided into one-hour periods, and individual animal movement was drawn at 10 min intervals to see which sheep moved first. The individual or individuals moving first were visually identified as separate from the other animals by approximately 1 m distance. Animals were ranked based on their movement trajectory relative to group members by assigning the highest ranks to the animal (s) that moved first and the lowest ranks to the last animals to move (for example, rank 7 to animal F, and rank 1 to animal E in Figure 5a). Animals were ranked the same if they moved together (Figure 5b) or received a zero if they did not move (animal B, Figure 5c) across the 10 min time intervals. The rank of individual animals was calculated for each one-hour period for a total of four hours across the two study days. The 10 min values ($n = 24$ values/animal) were summed to provide a score for each animal. Individual ranks were then drawn in the group's social network using ggnet2 function ('Ggally' package in R, [40]) for each one-hour period for a total of four hours. To provide better estimates of individuals' movement leader scores in the network, bar charts of scores for individuals across each hour were also plotted.

2.4.3. Sampling Rate and Social Networks

To estimate the optimal sampling rate for detecting an individual's nearest neighbours, four sampling approaches were examined. These included recording intervals of (1) 5 s (1 sample every 5 s—the initial frequency was every second but it was impossible to compute continuous 1 s data due to the processing power of the PC), (2) 5 min (1 sample every 5 min), (3) 1 min of recording at 1 s intervals and 4 min off, and (4) 1 min of recording at 1 s intervals and 9 min off. These sampling intervals were tested as battery power would be minimally affected by reducing the sampling interval to every 10 s or every 30 s, however, 4 min off would extend the battery life by a factor of 5. The data collected during the second day of the study (a total of 6 h and 30,240 observations based on a continuous sampling rate of 5 s) were used for this analysis. As a small number of individuals were used ($n = 7$), it was assumed that all the animals would be close enough to each other at least once to capture all the neighbours for each individual. The nearest neighbours for all animals were detected using the function of edge_nn of spatsoc package [29] while considering three different threshold distances at a maximum of: 10 cm, 20 cm, and 30 cm apart. The number of neighbours for each animal was then counted and plotted based on the different sampling rates and threshold distances using ggplot2 [41] in R [40].

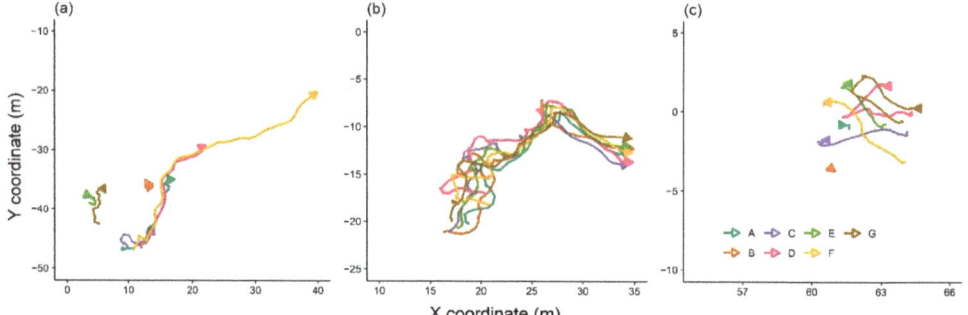

Figure 5. Some examples of individual animal movement behaviour at 10 min intervals on the second day of the study to rank the animals based on their movement trajectories and select the animal (s) who moved first. Plot (**a**) shows where all individuals were assigned a different rank, (**b**) shows where all individuals received the same rank, (**c**) shows where one animal (B) did not move. Individual animals are represented by separate colours and letters (A to G), and the direction of the arrow indicates the direction of travel at the conclusion of the 10 min period. The length of the lines indicates the distance travelled. For clarity, the specific paddock area the animals were within is displayed, hence the different axis values for each plot.

2.4.4. Social Network Comparison between Recorded Video and GNSS Data

To determine the accuracy of the RTK rovers for studying social networks in sheep, the position of individuals relative to each other were examined based on (1) recorded videos by the installed camera at the north corner of the paddock (Figure 6), (2) social network analyses of GNSS locational data, and (3) plotted GNSS location of animals. For this purpose, four 30 s time periods during the third day of the study were selected based on the criteria of good performance of all sensors, videos of high enough resolution to distinguish individual animal's positions relative to each other, and minimal movement by the animals. Static images at the beginning of the 30 s periods were extracted from the video recordings and were compared with the social network graph and GNSS positional plot. The distance between individuals was calculated using the spatsoc package [29]. The social network graph was then drawn using the package of ggplot2 [41] within the 30 s time periods with a threshold distance of 30 cm. The width of the edges in the graph was set based on the distance between individuals so that the width increased as the distance between dyads increased. The location of individuals was also plotted based on GNSS positional data using the package of ggplot2 [41].

3. Results

3.1. Device Accuracy of Laboratory Tests

The preliminary testing of the system performance showed that the antennae worked at a range of 950 m, and that even with misalignment, the system worked with RTK corrections out to 450 m. At the initial (non-RTK) GNSS lock, the errors were of the order of 5 to 10 m. The results improved to approximately 2 to 3 metres as differential GNSS became available (Figure 7a). The initial RTK lock is relatively difficult to acquire for moving sensors but can be improved by starting with remembering the last measured location, and a full RTK lock was only achieved after the placement on the stands (Figure 7b). The regularity of the sensor plots (Figure 7b,c) indicates that for the areas with an unobstructed LOS, the relative errors drop below 0.1 m. As the grid frame was walked (by researchers) into an area of tree blockage (Figure 7d), and some loss of the GNSS signal at the rovers occurred, the regularity of the array decreased showing errors around 1 m. That level of error was also shown when the experimenters' bodies blocked the GNSS signal to the satellites or the correction signal from the base station. The signal blockage was less of a problem during the farm trial. The GNSS antennas for the rovers were fitted on the back of the sheep and when the sheep were standing, the rover had a hemispheric sky view. Some blockage of the GNSS signal occurred when the sheep lay down, when the harness slipped to one side of the sheep, and from corner sections of the field fence line where solid metal railings were installed. The correction signal from the base station to the rover was minimally degraded during the test days, however, was blocked when people or vehicles were in the field during the preliminary validation stages.

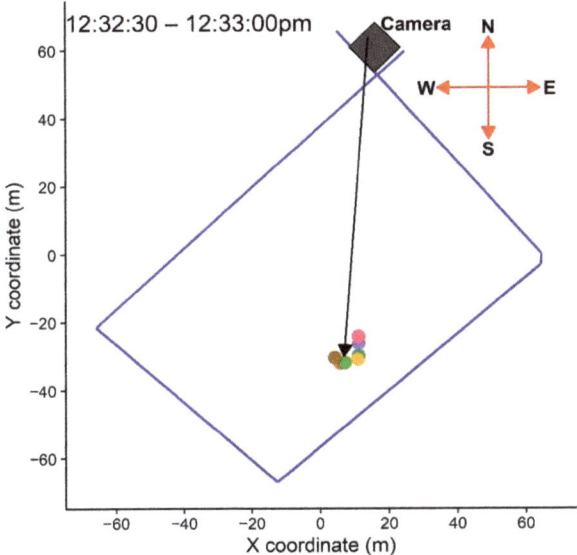

Figure 6. Position of the camera outside of the paddock (the north corner) to record the animals' movements during parts of the experiment to match the animal position with locational plots and social network graphs (see Section 3.3). Note: The dot points in the plot show the individuals within a 30 s time-period beginning at 12:32:30 p.m.

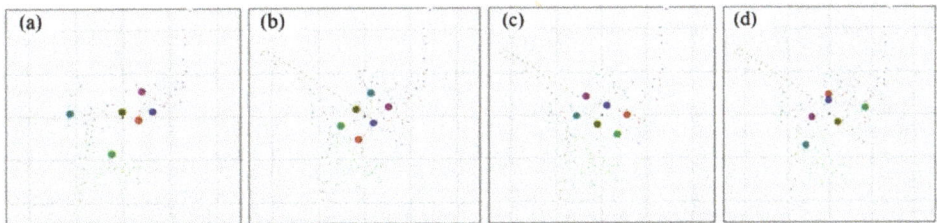

Figure 7. Results of laboratory testing the RTK rovers before placement on animals. Each figure shows the measured location of six rovers that were rigidly fixed on a 1 m × 2 m grid. (**a**) Before RTK fix was achieved and showing differential GNSS accuracy. (**b**) Rovers stationary with a clear view of the sky. (**c**) Rovers moving with a clear view of the sky. (**d**) Rovers moving underneath tree foliage and losing view of GNSS satellites and during the walk behind the trees which led to blocking of the satellite signals and the RTK correction link (heading NE). Note: The grid size is 2 m and the centre of the plot tracks to the centre of the array of RTK rovers. The large dots indicate the current location of the RTK rovers with the smaller dots displaying their movement history. Plots b and c show the situations similar to that expected in the real trial in the field.

The RTK rover errors were of the order of 0.1 m when LOS was available, and the system still showed significant enhancement over standard GNSS in areas with tree coverage and limited LOS to the base station. An advantage of the system is that the recorded data stream contains estimates of the quality of the location estimate, which identify areas where the location may have problems, and should thus be excluded from the analysis.

3.2. Device Accuracy and Reliability of Data Collection during Field Implementation on Sheep

For an estimate of the accuracy of the device when placed on animals, the paddock fence lines drawn based on GNSS positional data were placed on the top of the fitted line as shown in Figure 8a which indicates the high accuracy of the GNSS device in the paddock. The measurement errors orthogonal to the fence lines varied from 0 m to a maximum of 0.25 m for the north-west fence line due to the large and complex gate structure which made the line harder to estimate (Figure 8b). In addition to accuracy, the reliability of continuous data collection is another criterion required from a GNSS device. Overall, the RTK rovers worked well and recorded across the study duration, except for a temporary recording issue with rover C on the first day. Rover G was not used on day 1 due to accidental damage, and rover F did not record on day 4 due to operator error. Of 28 total sampling days (7 rovers × 4 study days), continuous data were obtained for 25 of these available (Figure 9).

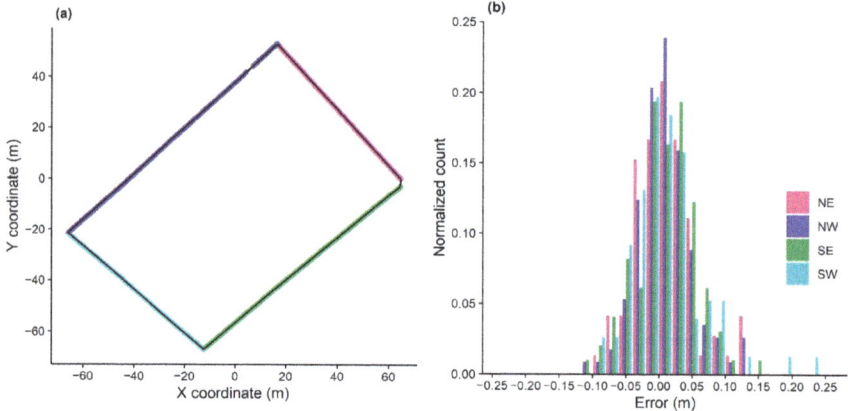

Figure 8. Graphs of (**a**) the fence line coordinates based on GNSS positional data with an optimally fit line (black line) and the GNSS positional points (coloured lines), and (**b**) the measurement errors of each fence line (north–east, north–west, south–east, and south–west). Note: To normalise the count, the frequency of observations in each bin was divided by the total number of observations.

3.3. Social Networks Based on Recorded Video and GNSS Data

Figure 10 presents the social networks of study animals based on recorded video (a), social network analysis (b), and GNSS positional data (c) at four time points on the third day of study. The video images (Figure 10a) confirm that the GNSS data recorded via the RTK rovers were able to correctly detect the position of animals relative to each other (Figure 10b,c). The relative positions of animals were detectable from the video recordings, but exact distances between animals could not be confirmed due to the front-on (cf. top-down) video records. The plotted relative position of animals based on GNSS positional data (Figure 10c) was similar to corresponding results from the social network analysis in which a thinner edge width indicates a closer distance (Figure 10b).

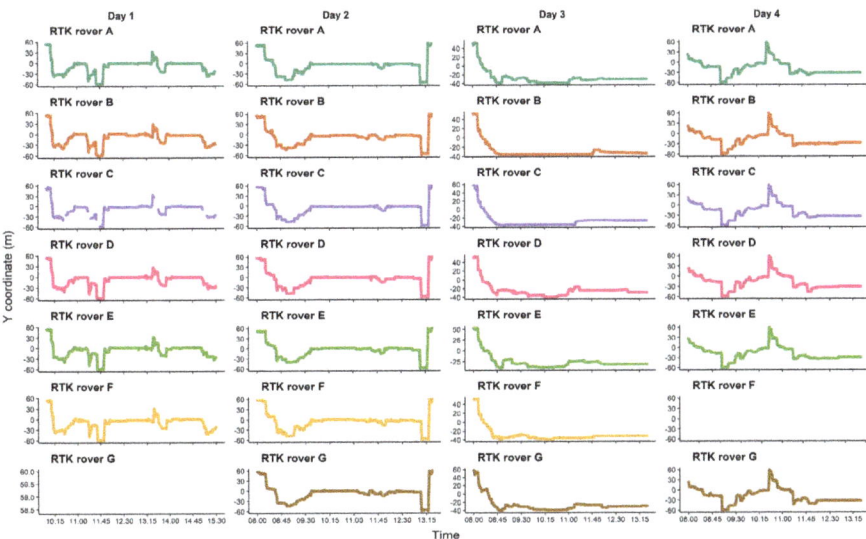

Figure 9. Consistency of recording for the RTK rovers on sheep over 5:30 h each across four study days. Due to some technical issues, RTK rover C did not record the positional data continuously on day 1 and RTK rover F did not work on day 4 (operator error). RTK rover G was not used on day 1 due to physical damage from transport.

3.4. Identifying Leaders from Movement Patterns

Figure 11a–e presents the overall movement leader scores during grazing movement as well as the pattern of change over four-hours on days 2 and 3 (two-hours each day). Overall, animal F had the highest movement leader score in the group as well as for two other hours (the second hour of day 2 (Figure 11c) and the first hour of day 3 (Figure 11d) while animal B had the lowest overall movement rank (Figure 11a). The bar charts present the movement leader scores across 10 min intervals per hour in which animals were assigned a number from 7 to 1 based on the order of movement (first to last respectively). Animals were assigned the same rank if they moved together and received a zero if they did not move. Consequently, not all 10 min intervals had an animal assigned as number 7 where if two animals moved first together, they both received a value of 6 instead. As the figure shows, some individuals were more likely to be first to move in the group. For example, animal E during the last two 10 min periods (41–50, and 51–60) of the first hour on day 2 (Figure 11b) or animal F during the last 10 min of third hour on day 3 (Figure 11d), but this was inconsistent across time and typically several individuals were identified as moving first together (e.g., animals E and F during the last 10 min of the second hour on day 2, Figure 11c) or sometimes all the animals moved together (e.g., the first 10 min of the fourth hour on day 3, Figure 11e).

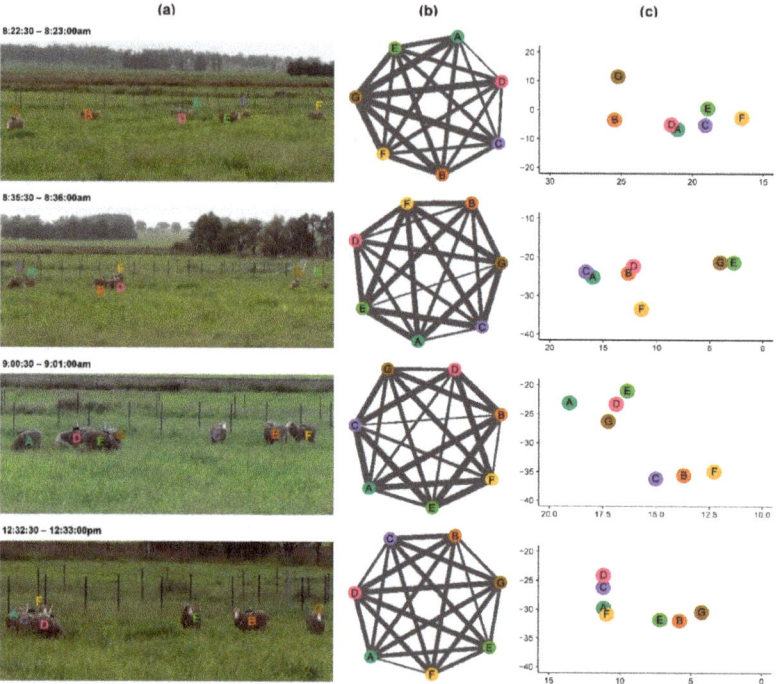

Figure 10. Animal social groups based on (**a**) recorded video (**b**), social network analysis, and (**c**) GNSS positional data at four time points on the third day of the study. In the social network plots, letters in the vertices (nodes) refer to individual animals and edges show the distance between two animals where a thinner line corresponds to a shorter distance and vice versa. Note: For each picture, the related social network graph and GNSS locational plot are in the same row.

3.5. Sampling Frequency and Social Network Analysis

As shown in Figure 12, the number of neighbours for each individual decreased when the sampling rate changed from continuous 5 s sampling to single records within 5-min time intervals (Figure 12a–c). However, when the interval of recording was changed to 1 min recording and 4 min off or even 9 min off, all neighbours were captured for the individuals with a threshold distance of 20 cm or 30 cm (Figure 12b,c). In contrast, with a sampling rate of 5 min, it was not possible to detect the neighbours for some time points even with a threshold distance of 30 cm (Figure 12c) indicating some or all social interactions were missed between individuals.

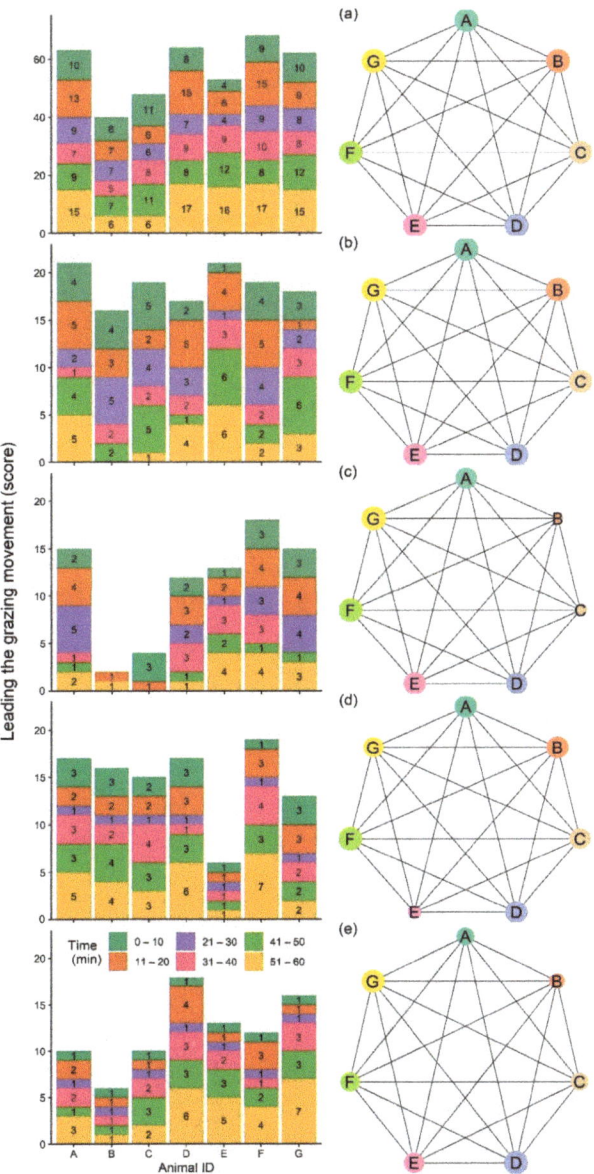

Figure 11. The pattern over time of individual gazing movement leader scores (**a**) summed for the total four-hour period during days 2 and 3; (**b**,**c**) first and second hour on day 2, and (**d**,**e**) first and second hour on day 3 presented as bar charts showing the individual movement rank for each 10 min period per hour of study (**b**–**e**) or the total score for 4 study hours (**a**) in which the animal (s) with the highest movement score moved first within each specific time point. A corresponding social network for each bar chart is also displayed (the larger vertex corresponds with an individual's higher rank). Letters in the vertices refer to individual animals.

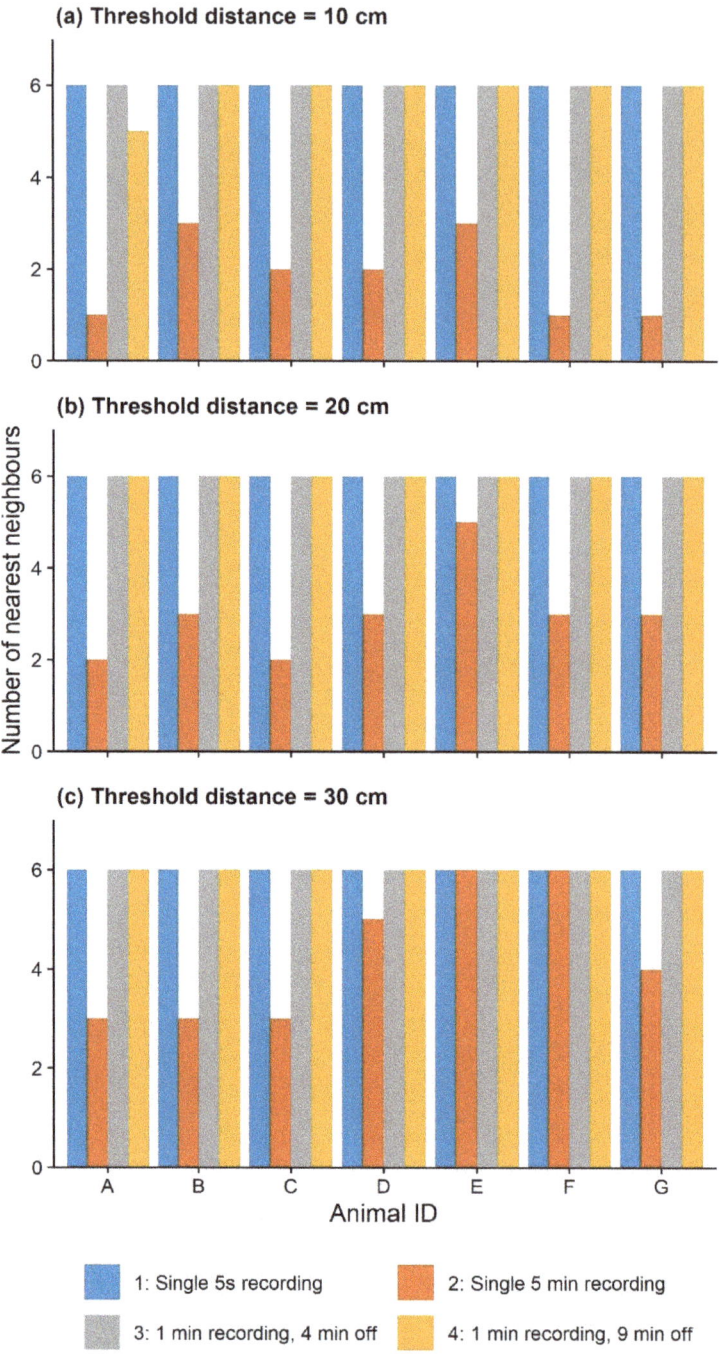

Figure 12. Change in the number of nearest neighbours for individual sheep (A–G) based on different sampling rates (single 5 s, single 5 min, 1 min on (at 1 s) and 4 min off, and 1 min on (at 1 s) and 9 min off) and threshold distances: (**a**) 10 cm, (**b**) 20 cm, and (**c**) 30 cm, during day 2 of the study.

4. Discussion

This study aimed to investigate the performance of RTK rovers that were developed to study movement leaders and social networks in seven sheep with high geo-spatial accuracy across four days. The devices were able to record while attached to sheep's backs and provide positional data with a relative error of approximately 0.2 m. These data showed that across a 4-h period there were no consistent individuals that always initiated movements of the group, but some individuals did move first more often than others. The number of detected neighbours was dependent on the sampling rate and threshold detection distance with 1 min recording and 4 or even 9 min off showing the same number of neighbours as continuous 5 s sampling at a threshold distance of 20 or 30 cm. RTK rovers may be a precision monitoring tool for greater understanding of sheep social behaviour but further work would be needed to improve the robustness of the devices for application on livestock.

When the RTK rovers were assessed for accuracy of recording positional data in the laboratory, the preliminary testing showed the aligned antennae performance was 950 m, and even with misalignment, the system functioned out to 450 m with the range theoretically considered to be several kilometres (but not tested within this study). With unobstructed line-of-sight (LOS) such as those conditions in the grassed field, the measurement errors were below 0.1 m. However, this error was affected by blockages and thus measuring animals in more complex terrain with trees and/or rocks, would reduce the recording accuracy. Additionally, the animals' bodies could also obstruct signals, particularly if a device slipped down which would be problematic for long-term deployment. This study was the first application of these specific devices on sheep and, thus, further testing and development of the RTK rovers would be needed for robust application onto animals.

From the positional data, it was possible to identify the sheep that first moved during a grazing period. The results indicated that some individuals within the group were more likely to be identified as initiating movement away from the group first, but this was not consistent across time, often several individuals moved first together, not all other individuals followed, and all individuals led at least one grazing movement across the 4 selected hours of observation. Leadership and movement hierarchies have previously been demonstrated in sheep with different types of transitional behaviours such as movement onto the pasture after a resting period [42], movement to a shearing shed [5,43], entering a raceway for weighing or leaving feed to rest under trees [44]. The identified leader, consistency in leader individuals, or whether others may follow a certain leader or not can also vary depending on the activity being observed [5,44], the previously established social bonds between individuals [42] or sheep breed [5,45]. Leadership during actual grazing may be less prominent than leadership during movement to new areas (for grazing or water), or leadership during grazing initiation [46]. Given that the current study was limited to a single group of 7 animals across a 4 h observation period, the conclusions that were able to be drawn regarding leadership in sheep within this study context are limited. However, the data generated by the RTK rovers were able to identify individual (s) that moved ahead of the group and, thus, the devices could be applied in future studies to understand what individual-level factors (e.g., temperament) affect leadership, and how leadership may change under varying circumstances (e.g., grazing, behavioural transitions, movement to new areas) or external conditions (e.g., weather, time of day). The development of automated algorithms that would detect the first animal (s) to move away from the group would further streamline the process and greatly enhance the information able to be gained from these types of large datasets.

The GNSS positional data were demonstrated to accurately quantify social networks in the studied group of sheep which was confirmed with recorded videos. Previous studies have also used on-animal devices to determine association patterns when sheep are first mixed together [36], flock level patterns when a 'predator' (herding dog) is presented to a group [4], and distance from peers either pre- or post-lambing [47]. Thus, the applications for social network analysis for sheep (and other livestock) are extensive with the technology having the potential to further our understanding of sheep behaviour as well as identify strategies to improve management practices that will enhance animal welfare and production. A technology with the precision of the RTK rovers in this study could increase the scope of questions able to be answered with more guaranteed accuracy.

Importantly, for sensor devices to be applied in future studies, the practical limitations of precision technology need to be considered. For automated on-animal devices, there is often a trade-off between high sampling frequency for a complete dataset and battery life. In this study, an intermittent sampling period was as informative as continuous 5 s sampling, if it was a continuous period. A continuous one min period with breaks of 4 or 9 min was adequate, but intermittent 5 min sampling was not, indicating dynamic positional data were needed to accurately quantify the networks. Similarly, Perreault, (2010) [39] modelled variations of social network data and demonstrated how an incomplete sample size results in incorrect network properties. Missing interactions may falsify the conclusions that are drawn from social network analysis [39]. In our analyses, the reduced sampling frequency provided the possibility of extending battery life but still generated sufficient data for accurate social network analysis at two relatively small threshold distances. However, this may not work in all research situations. While 4 min off may save 80% of the battery power, it does take up to 1 min for the devices to power up and lock onto the satellites. Thus, battery saving strategies would result in loss of data continuity. In this study of sheep grazing in a small paddock, loss of data continuity still produced the same social network results, but in situations where rapid social responses need to be observed (e.g., group responses to an active threat: King et al., 2012 [4]) loss of data continuity would likely compromise the social behaviour interpretations. There may also be a balance between sampling frequency and threshold distance. The greater threshold distances (20 or 30 cm) in this study allowed for intermittent data sampling in comparison with a 10 cm threshold distance. Haddadi et al. (2011 [36]) reported similar results where the most accurate social network data were a balance between sampling frequency and threshold distance with the most optimal sampling regime being 2.5 m distance for a continuous 3 min period. The selected sampling may also vary depending on the activity state of the animal group (e.g., moving as a flock to a new field, versus settled in the new field). The threshold distances in this study were smaller threshold distances than those tested by Haddadi et al. [36], 2–3 m and those typically set for proximity logger detection of social interactions in groups of sheep (e.g., 1–1.5 m [10,48]). The precision of the RTK rovers may be beneficial or necessary for specific close contact situations such as ewe-lamb relationships [25].

For long-term deployment of GNSS-enabled devices to accurately detect social networks, several options are possible to extend battery life. These include intermittent sampling by turning the GNSS on and off on a regular schedule such as one min on and 9 min off, synchronised across all individuals. This is likely to be advantageous only if the social network evolves significantly across one minute during periods of rapid movement. Alternatively, the GPS could be turned off until an accelerometer detects movement. This measurement may not be synchronised across the sheep and will suffer from errors in the inertial position calculation. Another option is to operate the GNSS in a standard 3D fix mode (with 10 m accuracy) and only use RTK-GNSS if some trigger event occurs such as a predator threat, rapid change in weather, or management intervention. Further application and testing under the different options presented above are needed to validate the optimal device design and data collection approach under long term studies.

5. Conclusions

The results of this study demonstrated high accuracy and reliable data recording of RTK rovers to measure social networks in sheep. From the positional data, it was possible to identify sheep that were more likely to move first, but no single individuals were consistently identified. The accuracy of the RTK rovers, however, provide the capability to identify leader animals across different scenarios such as movement towards new feed areas or a water supply, or when transitioning between behavioural states such as resting and grazing. Social networks of the group were generated with high accuracy using an intermittent sampling frequency, opening up the potential for battery saving and deployment in longer-term studies on livestock in the field. However, this was only a single group of sheep in an initial test of the RTK rovers and thus many further opportunities exist to apply the devices to more groups across different contexts to further understand social networks in sheep. Future work could use this device to investigate (1) the interaction between livestock and their environment, (2) how sheep with different personality traits behave in a social network, and (3) how regrouping animals might influence leadership and association patterns among individuals.

6. Patents

There is no patent resulting from the work reported in this manuscript.

Author Contributions: H.K., W.N., D.L.M.C. and C.L. contributed conception and design of the study. H.K., M.J. and D.A. conducted the animal experiment. H.K. and M.J. performed the statistical analysis. H.K. wrote the first draft of the manuscript. H.K., M.J., D.L.M.C. and C.L. wrote sections of the manuscript. All authors contributed to manuscript revision, read, and approved the submitted version. All authors have read and agreed to the published version of the manuscript.

Funding: This work was funded by the Commonwealth Scientific and Industrial Research Organisation (CSIRO) (internal funding, URL: www.csiro.au) and "HK was supported by a CSIRO Research Plus postdoctoral fellowship".

Institutional Review Board Statement: The study was conducted according to the guidelines of the Declaration of Helsinki, and approved by the Institutional Animal Ethics Committee of the CSIRO FD McMaster Laboratory Chiswick (ARA 19-27 approved 19 December 2019)."

Informed Consent Statement: Not applicable.

Data Availability Statement: Data supporting this study will be made available upon any reasonable request to the corresponding author.

Acknowledgments: The authors are grateful to Troy Kalinowski (CSIRO), Tim Dyall (CSIRO), and Jim Lea (CSIRO) for technical support, and Alec L. Robitaille (Department of Biology, Memorial University of Newfoundland, Canada) for helping with the social network analysis.

Conflicts of Interest: There is no conflict of interest.

References

1. Kendrick, K.M.; da Costa, A.P.; Leigh, A.E.; Hinton, M.R.; Peirce, J.W. Sheep Don't Forget a Face. *Nature* **2001**, *414*, 165–166. [CrossRef] [PubMed]
2. Tate, A.J.; Fischer, H.; Leigh, A.E.; Kendrick, K.M. Behavioural and Neurophysiological Evidence for Face Identity and Face Emotion Processing in Animals. *Philos. Trans. R. Soc. B Biol. Sci.* **2006**, *361*, 2155–2172. [CrossRef] [PubMed]
3. Fisher, A.; Matthews, L. The social behaviour of sheep. In *Social Behaviour in Farm Animals*; Keeling, L.J., Gonyou, H.W., Eds.; CABI: Wallingford, CT, USA, 2001; pp. 211–245. ISBN 978-0-85199-397-3.
4. King, A.J.; Wilson, A.M.; Wilshin, S.D.; Lowe, J.; Haddadi, H.; Hailes, S.; Morton, A.J. Selfish-Herd Behaviour of Sheep under Threat. *Curr. Biol.* **2012**, *22*, R561–R562. [CrossRef] [PubMed]
5. Ortiz-Plata, C.; De Lucas-Tron, J.; Miranda-de la Lama, G.C. Breed Identity and Leadership in a Mixed Flock of Sheep. *J. Vet. Behav.* **2012**, *7*, 94–98. [CrossRef]
6. Boissy, A.; Dumont, B. Interactions between Social and Feeding Motivations on the Grazing Behaviour of Herbivores: Sheep More Easily Split into Subgroups with Familiar Peers. *Appl. Anim. Behav. Sci.* **2002**, *79*, 233–245. [CrossRef]
7. Sibbald, A.M.; Hooper, R.J. Trade-Offs between Social Behaviour and Foraging by Sheep in Heterogeneous Pastures. *Behav. Process.* **2003**, *61*, 1–12. [CrossRef]
8. Dumont, B.; Boissy, A. Grazing Behaviour of Sheep in a Situation of Conflict between Feeding and Social Motivations. *Behav. Process.* **2000**, *49*, 131–138. [CrossRef]
9. Doyle, R.E.; Broster, J.C.; Barnes, K.; Browne, W.J. Temperament, Age and Weather Predict Social Interaction in the Sheep Flock. *Behav. Process.* **2016**, *131*, 53–58. [CrossRef]
10. Ozella, L.; Langford, J.; Gauvin, L.; Price, E.; Cattuto, C.; Croft, D.P. The Effect of Age, Environment and Management on Social Contact Patterns in Sheep. *Appl. Anim. Behav. Sci.* **2020**, *225*, 104964. [CrossRef]

11. Hauschildt, V.; Gerken, M. Individual Gregariousness Predicts Behavioural Synchronization in a Foraging Herbivore, the Sheep (*Ovis Aries*). *Behav. Process.* **2015**, *113*, 110–112. [CrossRef]
12. Gougoulis, D.A.; Kyriazakis, I.; Fthenakis, G.C. Diagnostic Significance of Behaviour Changes of Sheep: A Selected Review. *Small Rumin. Res.* **2010**, *92*, 52–56. [CrossRef]
13. Wurtz, K.; Camerlink, I.; D'Eath, R.B.; Fernández, A.P.; Norton, T.; Steibel, J.; Siegford, J. Recording Behaviour of Indoor-Housed Farm Animals Automatically Using Machine Vision Technology: A Systematic Review. *PLoS ONE* **2019**, *14*, e0226669. [CrossRef] [PubMed]
14. Dumont, B.; Boissy, A.; Achard, C.; Sibbald, A.M.; Erhard, H.W. Consistency of Animal Order in Spontaneous Group Movements Allows the Measurement of Leadership in a Group of Grazing Heifers. *Appl. Anim. Behav. Sci.* **2005**, *95*, 55–66. [CrossRef]
15. Bailey, D.W.; Trotter, M.G.; Knight, C.W.; Thomas, M.G. Use of GPS Tracking Collars and Accelerometers for Rangeland Livestock Production Research. *Transl. Anim. Sci.* **2018**, *2*, 81–88. [CrossRef] [PubMed]
16. Drewe, J.A.; Weber, N.; Carter, S.P.; Bearhop, S.; Harrison, X.A.; Dall, S.R.X.; McDonald, R.A.; Delahay, R.J. Performance of Proximity Loggers in Recording Intra- and Inter-Species Interactions: A Laboratory and Field-Based Validation Study. *PLoS ONE* **2012**, *7*, e39068. [CrossRef]
17. Rocha, L.E.C.; Terenius, O.; Veissier, I.; Meunier, B.; Nielsen, P.P. Persistence of Sociality in Group Dynamics of Dairy Cattle. *Appl. Anim. Behav. Sci.* **2020**, *223*, 104921. [CrossRef]
18. Ren, K.; Karlsson, J.; Liuska, M.; Hartikainen, M.; Hansen, I.; Jørgensen, G.H. A Sensor-Fusion-System for Tracking Sheep Location and Behaviour. *Int. J. Distrib. Sens. Netw.* **2020**, *16*, 155014772092177. [CrossRef]
19. Boyland, N.K.; Mlynski, D.T.; James, R.; Brent, L.J.N.; Croft, D.P. The Social Network Structure of a Dynamic Group of Dairy Cows: From Individual to Group Level Patterns. *Appl. Anim. Behav. Sci.* **2016**, *174*, 1–10. [CrossRef]
20. Xu, H.; Li, S.; Lee, C.; Ni, W.; Abbott, D.; Johnson, M.; Lea, J.M.; Yuan, J.; Campbell, D.L.M. Analysis of Cattle Social Transitional Behaviour: Attraction and Repulsion. *Sensors* **2020**, *20*, 5340. [CrossRef]
21. Williams, B. Resource Selection by Hill Sheep: Direct Flock Observations versus GPS Tracking. *Appl. Ecol. Environ. Res.* **2010**, *8*, 279–299. [CrossRef]
22. Keshavarzi, H.; Lee, C.; Lea, J.M.; Campbell, D.L.M. Virtual Fence Responses Are Socially Facilitated in Beef Cattle. *Front. Vet. Sci.* **2020**, *7*, 543158. [CrossRef] [PubMed]
23. O'Neill, C.J.; Bishop-Hurley, G.J.; Williams, P.J.; Reid, D.J.; Swain, D.L. Using UHF Proximity Loggers to Quantify Male–Female Interactions: A Scoping Study of Estrous Activity in Cattle. *Anim. Reprod. Sci.* **2014**, *151*, 1–8. [CrossRef] [PubMed]
24. Paganoni, B.; Macleay, C.; van Burgel, A.; Thompson, A. Proximity Sensors Fitted to Ewes and Rams during Joining Can Indicate the Birth Date of Lambs. *Comput. Electron. Agric.* **2020**, *170*, 105249. [CrossRef]
25. Sohi, R.; Trompf, J.; Marriott, H.; Bervan, A.; Godoy, B.I.; Weerasinghe, M.; Desai, A.; Jois, M. Determination of Maternal Pedigree and Ewe–Lamb Spatial Relationships by Application of Bluetooth Technology in Extensive Farming Systems. *J. Anim. Sci.* **2017**, *95*, 5145–5150. [CrossRef]
26. Buerkert, A.; Schlecht, E. Performance of Three GPS Collars to Monitor Goats' Grazing Itineraries on Mountain Pastures. *Comput. Electron. Agric.* **2009**, *65*, 85–92. [CrossRef]
27. Fogarty, E.S.; Swain, D.L.; Cronin, G.; Trotter, M. Autonomous On-Animal Sensors in Sheep Research: A Systematic Review. *Comput. Electron. Agric.* **2018**, *150*, 245–256. [CrossRef]
28. Farine, D.R. A Guide to Null Models for Animal Social Network Analysis. *Methods Ecol. Evol.* **2017**, *8*, 1309–1320. [CrossRef]
29. Robitaille, A.L.; Webber, Q.M.R.; Vander Wal, E. Conducting Social Network Analysis with Animal Telemetry Data: Applications and Methods Using Spatsoc. *Methods Ecol. Evol.* **2019**, *10*, 1203–1211. [CrossRef]
30. Šárová, R.; Špinka, M.; Panamá, J.L.A.; Šimeček, P. Graded Leadership by Dominant Animals in a Herd of Female Beef Cattle on Pasture. *Anim. Behav.* **2010**, *79*, 1037–1045. [CrossRef]
31. Guo, Y.; Poulton, G.; Corke, P.; Bishop-Hurley, G.J.; Wark, T.; Swain, D.L. Using Accelerometer, High Sample Rate GPS and Magnetometer Data to Develop a Cattle Movement and Behaviour Model. *Ecol. Model.* **2009**, *220*, 2068–2075. [CrossRef]
32. Webber, B.; Weber, K.; Clark, P.; Moffet, C.; Ames, D.; Taylor, J.; Johnson, D.; Kie, J. Movements of Domestic Sheep in the Presence of Livestock Guardian Dogs. *Sheep Goat Res. J.* **2015**, *30*, 18–23.
33. Ganskopp, D.C.; Johnson, D.D. GPS Error in Studies Addressing Animal Movements and Activities. *Rangel. Ecol. Manag.* **2007**, *60*, 350–358. [CrossRef]

34. Davis, M.J.; Thokala, S.; Xing, X.; Hobbs, N.T.; Miller, M.W.; Han, R.; Mishra, S. Testing the Functionality and Contact Error of a GPS-Based Wildlife Tracking Network. *Wildl. Soc. Bull.* **2013**, *37*, 855–861. [CrossRef]
35. Li, X.; Zhang, X.; Ren, X.; Fritsche, M.; Wickert, J.; Schuh, H. Precise Positioning with Current Multi-Constellation Global Navigation Satellite Systems: GPS, GLONASS, Galileo and BeiDou. *Sci. Rep.* **2015**, *5*, 8328. [CrossRef] [PubMed]
36. Haddadi, H.; King, A.J.; Wills, A.P.; Fay, D.; Lowe, J.; Morton, A.J.; Hailes, S.; Wilson, A.M. Determining Association Networks in Social Animals: Choosing Spatial–Temporal Criteria and Sampling Rates. *Behav. Ecol. Sociobiol.* **2011**, *65*, 1659–1668. [CrossRef]
37. Catania, P.; Comparetti, A.; Febo, P.; Morello, G.; Orlando, S.; Roma, E.; Vallone, M. Positioning Accuracy Comparison of GNSS Receivers Used for Mapping and Guidance of Agricultural Machines. *Agronomy* **2020**, *10*, 924. [CrossRef]
38. Farine, D.R.; Whitehead, H. Constructing, Conducting and Interpreting Animal Social Network Analysis. *J. Anim. Ecol.* **2015**, *84*, 1144–1163. [CrossRef]
39. Perreault, C. A Note on Reconstructing Animal Social Networks from Independent Small-Group Observations. *Anim. Behav.* **2010**, *80*, 551–562. [CrossRef]
40. R Core Team. *R: A Language and Environment for Statistical Computing*; R Foundation for Statistical Computing: Vienna, Austria, 2018.
41. Wickham, H. *Ggplot2: Elegant Graphics for Data Analysis*; Springer: New York, NY, USA; ISBN 978-3-319-24277-4. Available online: https://ggplot2.tidyverse.org,2016 (accessed on 17 July 2020).
42. Ramseyer, A.; Boissy, A.; Thierry, B.; Dumont, B. Individual and Social Determinants of Spontaneous Group Movements in Cattle and Sheep. *Animal* **2009**, *3*, 1319–1326. [CrossRef]
43. Sherwin, C.M.; Johnson, K.G. The Influence of Social Factors on the Use of Shade by Sheep. *Appl. Anim. Behav. Sci.* **1987**, *18*, 143–155. [CrossRef]
44. Sherwin, C.M. Priority of Access to Limited Feed, Butting Hierarchy and Movement Order in a Large Group of Sheep. *Appl. Anim. Behav. Sci.* **1990**, *25*, 9–24. [CrossRef]
45. Squires, V.R.; Daws, G.T. Leadership and Dominance Relationships in Merino and Border Leicester Sheep. *Appl. Anim. Ethol.* **1975**, *1*, 263–274. [CrossRef]
46. Sato, S. Leadership during Actual Grazing in a Small Herd of Cattle. *Appl. Anim. Ethol.* **1982**, *8*, 53–65. [CrossRef]
47. Dobos, R.C.; Dickson, S.; Bailey, D.W.; Trotter, M.G. The Use of GNSS Technology to Identify Lambing Behaviour in Pregnant Grazing Merino Ewes. *Anim. Prod. Sci.* **2014**, *54*, 1722. [CrossRef]
48. Vander Wal, E.; Gagné-Delorme, A.; Festa-Bianchet, M.; Pelletier, F. Dyadic Associations and Individual Sociality in Bighorn Ewes. *Behav. Ecol.* **2016**, *27*, 560–566. [CrossRef]

Publisher's Note: MDPI stays neutral with regard to jurisdictional claims in published maps and institutional affiliations.

© 2021 by the authors. Licensee MDPI, Basel, Switzerland. This article is an open access article distributed under the terms and conditions of the Creative Commons Attribution (CC BY) license (http://creativecommons.org/licenses/by/4.0/).

Article

Application of Plasma-Printed Paper-Based SERS Substrate for Cocaine Detection

Rhiannon Alder [1,2,†], Jungmi Hong [3,*,†], Edith Chow [3], Jinghua Fang [4], Fabio Isa [3], Bryony Ashford [3], Christophe Comte [3], Avi Bendavid [3], Linda Xiao [1], Kostya (Ken) Ostrikov [5], Shanlin Fu [1,2] and Anthony B. Murphy [3]

1. Centre for Forensic Science, University of Technology Sydney, Sydney, NSW 2007, Australia; Rhiannon.L.Alder@student.uts.edu.au (R.A.); Linda.Xiao@uts.edu.au (L.X.); Shanlin.Fu@uts.edu.au (S.F.)
2. IDEAL ARC Research Hub, University of Technology Sydney, Sydney, NSW 2007, Australia
3. CSIRO Manufacturing, Lindfield, NSW 2070, Australia; edith.chow@csiro.au (E.C.); fabio.isa@csiro.au (F.I.); bryony.ashford@csiro.au (B.A.); christophe.comte@csiro.au (C.C.); avi.bendavid@csiro.au (A.B.); tony.murphy@csiro.au (A.B.M.)
4. Aloxitec Pty Ltd., Lindfield, NSW 2070, Australia; aloxitec@gmail.com
5. School of Chemistry, Physics and Mechanical Engineering, Queensland University of Technology, Brisbane, QLD 4001, Australia; kostya.ostrikov@qut.edu.au
* Correspondence: jungmi.hong@csiro.au
† These authors contributed equally.

Academic Editor: Anna Chiara De Luca

Received: 29 December 2020; Accepted: 22 January 2021; Published: 26 January 2021

Abstract: Surface-enhanced Raman spectroscopy (SERS) technology is an attractive method for the prompt and accurate on-site screening of illicit drugs. As portable Raman systems are available for on-site screening, the readiness of SERS technology for sensing applications is predominantly dependent on the accuracy, stability and cost-effectiveness of the SERS strip. An atmospheric-pressure plasma-assisted chemical deposition process that can deposit an even distribution of nanogold particles in a one-step process has been developed. The process was used to print a nanogold film on a paper-based substrate using a $HAuCl_4$ solution precursor. X-ray photoelectron spectroscopy (XPS) analysis demonstrates that the gold has been fully reduced and that subsequent plasma post-treatment decreases the carbon content of the film. Results for cocaine detection using this substrate were compared with two commercial SERS substrates, one based on nanogold on paper and the currently available best commercial SERS substrate based on an Ag pillar structure. A larger number of bands associated with cocaine was detected using the plasma-printed substrate than the commercial substrates across a range of cocaine concentrations from 1 to 5000 ng/mL. A detection limit as low as 1 ng/mL cocaine with high spatial uniformity was demonstrated with the plasma-printed substrate. It is shown that the plasma-printed substrate can be produced at a much lower cost than the price of the commercial substrate.

Keywords: cocaine detection; plasma printing; SERS; gold nanoparticles; forensics; illicit drugs; on-site testing; paper substrate

1. Introduction

With the availability of portable Raman systems, there is an enormous opportunity to create low-cost, highly sensitive and reliable surface-enhanced Raman spectroscopy (SERS) strips for on-site testing of trace illicit drugs [1–5] and explosives [6,7]. The inelastic scattering of photons by incident light can be used to determine the vibrational modes of molecules and thus provide a structural fingerprint of the molecule. Compared to conventional Raman techniques [8], the use of a roughened metal or metal nanoparticle surface in SERS enhances the Raman effect by typically 6–8 orders of magnitude [5] owing to localised surface plasmonic resonances around the surface protrusions or particles [9]. Fedick et al. developed an undergraduate experiment that incorporated measurements using a commercial silver-on-paper SERS substrate for the detection of heroin, fentanyl and 3,4-methylenedioxymethamphetamine [10]. Inkjet printing methods have allowed the construction of SERS substrates that can be used for two-dimensional chromatographic separation for complex matrix analysis [11]. These inkjet-printed substrates allowed quantification of 25 ng of heroin mixed with highly fluorescent materials [11]. Silver nanoparticle-soaked filter paper discs were used by Haddad et al. for the analysis of fentanyl-spiked heroin [1]. The limit of detection for fentanyl was 100 ng/L when 10 µL of analyte solution was deposited. Swabbing with these substrates allowed the recovery of fentanyl from surfaces.

Despite the advances in SERS testing, there remains an opportunity to improve the analyte sensitivity and reduce the complexity in the fabrication of SERS-active substrates. Optimisation of the shape and size distribution of the nanometals allows enhancement of the Raman signal from analyte molecules. Sophisticated shapes and combinations of nanometal particles have been intensively studied, such as nanoflowers [12], nanostars [13], sea-urchin-shaped nanoparticles [14], as well as different types of core-shell structure [15]. Various fabrication techniques have also been applied, including electron beam lithography [16], nanosphere lithography [17] and focused ion beam patterning [18], to achieve high sensitivity and reproducibility. However, most are elaborate multi-step processes that are costly and are not amenable for large-scale production and on-site or point-of-care testing.

The choice of substrate and materials deposition technique are two other important considerations for SERS-based applications. Paper-based substrates are highly attractive as they are low-cost, disposable and can be readily modified by inkjet printing, drop-casting, direct writing and soaking with different nanomaterials [19–23]. Moreover, paper-based substrates are flexible, which could allow for swabbing applications [24,25] within the forensics field for detection of illicit drugs and explosives.

Inkjet printing [26–28] provides the best control over the uniformity of gold nanoparticle films with microscale precision and is amenable to high-throughput, rapid-prototyping of SERS substrates. However, the preparation of the ink requires several steps, including metal nanoparticle synthesis and filtering [29]; alternatively, commercial nanogold or nanosilver inks are available but costly. There is also the possibility of interference from residual chemicals such as reducing or stabilising agents in SERS measurements of low concentrations of analyte molecules.

Alternative methods that can reduce the fabrication complexity of nanogold substrates are highly desirable. Plasma deposition, especially atmospheric-pressure plasma-assisted chemical vapour deposition (CVD), has recently been demonstrated as a facile and cost-effective processing technique for nanogold deposition [30]. The advantage of this technique is that it is possible to produce and deposit nanogold films directly on the substrate from a single $HAuCl_4$ solution precursor without additional reducing or stabilising agents. Furthermore, there is no need for multi-step filtration, centrifugation or purification steps that are commonly employed in nanoparticle synthesis. A plasma jet of high-density electrons reduces the solution to metallic gold, which is then deposited on the substrate almost instantaneously. Since plasma printing does not require complex and multiple chemical steps, nor high-end electron or ion beam processing under vacuum, it is a promising alternative technique to fabricate highly sensitive, low-cost paper-based SERS substrates.

In this study, a paper-based SERS substrate is produced by depositing nano gold on paper using the plasma-assisted CVD technique. The substrate is applied to the detection of cocaine. Cocaine, an alkaloid compound, is the second most-consumed stimulant in Australia [31] with approximately 4.1 tonnes consumed each year [32,33].

The main routes of administration include insufflation, smoking and injection with the administration route depending on the drug form. In Australia, cocaine hydrochloride is the most common form [32] and is administered by insufflation or rubbing on the gums [34,35]. Consumers under the influence of cocaine can exhibit behaviour that is unpredictable, violent or aggressive, which can be dangerous to both themselves and others [34,36]. Current on-site testing for cocaine involves colour testing or immunoassay strips, which then require confirmatory testing [37]. Ideally, a method that can provide rapid on-site testing is highly desirable.

The plasma-printed SERS substrates are compared to two commercially available substrates, a paper-based gold SERS substrate and a silver pillar structure on Si that presents the state-of-the-art for cocaine detection. The plasma-printed SERS substrate shows promise as a simple and scalable fabrication technique for the highly sensitive detection of cocaine.

2. Materials and Methods

2.1. Chemicals

Whatman No. 1 filter paper and gold(III) chloride trihydrate ($HAuCl_4 \cdot 3H_2O$) were purchased from Sigma-Aldrich (Macquarie Park, Australia). Cocaine hydrochloride was purchased from the National Measurement Institute (West Lindfield, Australia). Ultra-pure Milli-Q water (>18.2 MΩ cm) and ethanol (Wilmar, Yarraville, Australia) were used for the preparation of solutions. Ar and He gases were purchased from BOC in 99.997% high purity and 99.999% ultra-high purity grades, respectively.

2.2. SERS Substrate Fabrication

Nanogold was deposited on paper (Whatman filter paper No. 1) from an $HAuCl_4$ precursor solution using the atmospheric-pressure plasma jet. 1% *w/v* $HAuCl_4$ aqueous solution was prepared and mixed with ethanol in 1:1 volume ratio to provide improved atomisation. Using a syringe pump (Harvard PHD 2000), 20 µL/min of the liquid source was supplied to a parallel-path pneumatic nebuliser (Burgener Research Inc., Ontario, Canada), which atomised the droplet into a fine vapour through interaction with a fast Ar gas stream as shown in Figure 1a,b. The plasma jet consisted of a custom-blown glass tube (Pyrex glass, inner diameter 6 mm, thickness 1.5 mm) and two parallel ring shape electrodes. It was powered by a high-voltage AC power supply (PVM500) operated typically at 23 kHz with a peak voltage 7.0 ± 0.5 kV. Using a mass flow controller (SevenStar D08), 0.5 LPM of Ar and 4 LPM of He were supplied to the active plasma discharge region. The plasma jet module is situated on a table-top CNC (Computer Numerical Control) machine (High Z—cncstep.de) in order to deposit and print a specified pattern. The separation between the glass tube aperture and the substrate was 2 mm.

The scanning motion of the plasma jet was tested with different scanning times, with parameters chosen to optimise the film properties for highly sensitive SERS measurement. In this work, all samples were deposited with six passes, each of 3 mm width at 1 mm/s. The influence of the number of passes is shown in Figure S1 in the SI.

Unless otherwise noted, the deposited films were plasma post-treated. This was done by scanning the films twice with the same plasma jet with He gas only at the same input power conditions (23 kHz, peak voltage 7.0 ± 0.5 kV) and the scanning speed of 1 mm/s.

2.3. Plasma Characterisation

The optical emission spectra were measured using an optical emission spectrometer (Acton SP2500/Princeton Instrument). The slit width was 10 µm, and the exposure time was 3 s. A fibre input coupler placed at a radial distance of 20 mm from the plasma jet was used. Measurements were performed for both the active plasma discharge region at the same height as the midpoint between the two electrodes, and at a position 2 mm above the substrate.

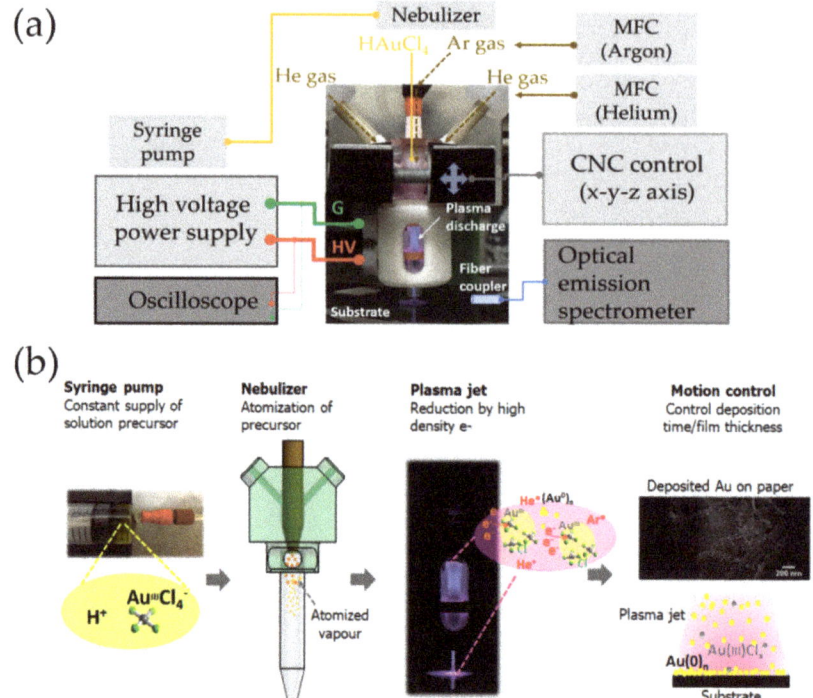

Figure 1. Experimental configuration (**a**) and illustration of the working principle and elements of the plasma jet printing process (**b**); adapted from Hong et al. [30] with permission from The Royal Society of Chemistry.

2.4. Surface Characterisation

XPS measurements were performed using a Specs150 SAGE instrument with an Mg Kα X-ray source with energy 1253.6 eV. The resolution for the energy scale is 0.1 eV and 15 scans are accumulated for the elemental analysis.

The surface morphology of the printed nanogold film were characterized using field-emission scanning electron microscope (FE-SEM.; Zeiss) operated at electron beam energies of 5 keV with an InLens secondary electron detector.

2.5. Standard Dilutions

A stock solution of cocaine was prepared by dissolving 1 mg of solid cocaine hydrochloride powder in 1 mL of MilliQ water. Serial dilutions were performed to produce standards with concentrations of 5000, 1000, 500, 100, 10 and 1 ng/mL. A 5 µL aliquot was deposited onto the SERS substrate [38].

2.6. Oral Fluid Extractions

Oral fluid was collected under human ethics approval No: UTS ETH18–2521. Oral fluid was spiked at concentrations of 10 and 100 ng/mL. Spiked and blank oral fluid samples (100 µL) were pH adjusted with 100 µL 0.1 M pH 9.2 carbonate buffer and extracted with 100 µL 9:1 dichloromethane (DCM): isopropyl alcohol (IPA) [38]. The extraction method was adapted from Clauwaert et al. [39]. A 5 µL aliquot of the organic phase was deposited onto the plasma-printed substrate.

2.7. Raman Analysis

Raman analysis was conducted using a Renishaw inVia Raman microscope with 785 nm laser and 1200 line/mm grating. The analysis was conducted with a laser power of 20 mW, over the range of 550–2000 cm^{-1} with 10 s exposure, single accumulation and pinhole in. The microscope objective was set to 20× magnification. For the detection of different concentrations of cocaine, 20 spectra were collected across different points on the substrate surface [38].

Raman mapping involved constructing a montage of the microscope images across the entire surface and taking consecutive measurements within a grid. The distribution of compounds on a surface was determined by picking characteristic bands and having the image displayed as an intensity heat map. Mapping was set up over the still image montage of the entire surface of the substrate. The montage was constructed using eight images in the x-direction and 13 images in the y-direction. Mapping steps were 175 μm × 175 μm in a grid for a total of 399 spectra collected across the surface. The mapping review was conducted using the intensity at a point across four common cocaine bands, 1003 cm^{-1}, 1032 cm^{-1}, 1450 cm^{-1} and 1600 cm^{-1}.

2.8. Comparison with Commercial SERS Substrates

Commercial gold paper-based SERS substrates (P-SERS) were purchased from Metrohm Australia Pty Ltd. (Sydney, Australia) and silver-coated silicon pillar substrates were purchased from JASMAT Optics Corp (Taiwan). Cocaine standard solutions ranging in concentration from 1–5000 ng/mL were used to compare the SERS spectra of the plasma-printed and commercial substrates. The visible cocaine vibrational bands were annotated on the spectra and tabulated.

3. Results

3.1. Plasma Characterisation

The estimated average power density in the plasma was 4.0 ± 0.3 W/cm^3, calculated using an estimated discharge volume of 1.6 cm^3. The average electron number density was calculated to be (1.4 ± 0.2) × 10^{10} cm^{-3}, as has been previously reported [30]. The optical emission spectrum was measured to understand the plasma reduction process in the active discharge zone between the electrodes and near the substrate. The emission spectra from various excited states of molecules, radicals, ions and atoms were observed, indicating a highly reactive environment of the plasma discharge with a low gas temperature of 360 ± 30 K. The detailed optical emission measurement results are provided in the SI. Unlike the high-temperature N$_2$ plasma with a large amount of chloroauric acid on the surface presented by Wu et al. [40], no excited AuCl molecules were observed. Maguire et al. [41] suggested the high density of electrons in the plasma discharge may provide a rapid reduction of HAuCl$_4$. Therefore, AuCl emission may not be observable because the lifetime of AuCl will be very short in a high-density plasma discharge with finely atomised vapour.

3.2. Surface Characterisation

SEM images, shown in Figure S2 in the SI, revealed uniform deposition of nanogold particles along the intrinsic matrix of the paper substrate. However, due to the charging problem, it was not possible to investigate the detailed structure and shape at high magnification. The performance of the deposited nanogold film as a SERS substrate was greatly improved by plasma post-treatment. Only plasma post-treated substrates were able to detect cocaine. They also showed significantly improved sensitivity in detecting low concentrations of Rhodamine B, as shown in Figures S3 and S4 in the SI. The improvement is attributed to the reduction of amorphous carbon content in the nanogold films. The carbon introduced by ethanol dissolved in the precursor solution. As we described in Section 2.2, the plasma post-treatment was done using a He plasma jet at atmospheric pressure. Because it is operated under ambient conditions, it can interact with the molecules in the surrounding air and generate reactive radicals. It is expected to introduce new functional groups and modify the surface properties, as is commonly reported for many plasma processes at atmospheric pressure. A decreased C-C bond and newly introduced oxygen functional groups are commonly observed when vacuum plasmas containing oxygen or ambient-air-exposed atmospheric-pressure plasmas are used to treat carbon-based organic materials such as fibres or polymeric substances [42,43]. Figure 2 shows the XPS spectra of Au4 f, C1s, O1s and Cl2p for the nanogold film before and after plasma post-treatment. The post-treatment causes a 0.4 eV shift in the Au4f peak and a clear increase in the O1 s peak intensity. The atomic composition of the nanogold film, obtained using SpecsLab analysis software, is given in Table 1, where the instrumental error, including variation of X-ray intensity, analyser pass energy, aperture settings, etc., is known to be at most. 1% for C or O, and is significantly lower for elements such as Au and Cl. The content of Cl was below the detection limit before and after post-treatment, indicating a high level of reduction of the ionic gold in the precursor. As shown in Figure 2, the oxygen content was increased, and the carbon content decreased, by the plasma post-treatment.

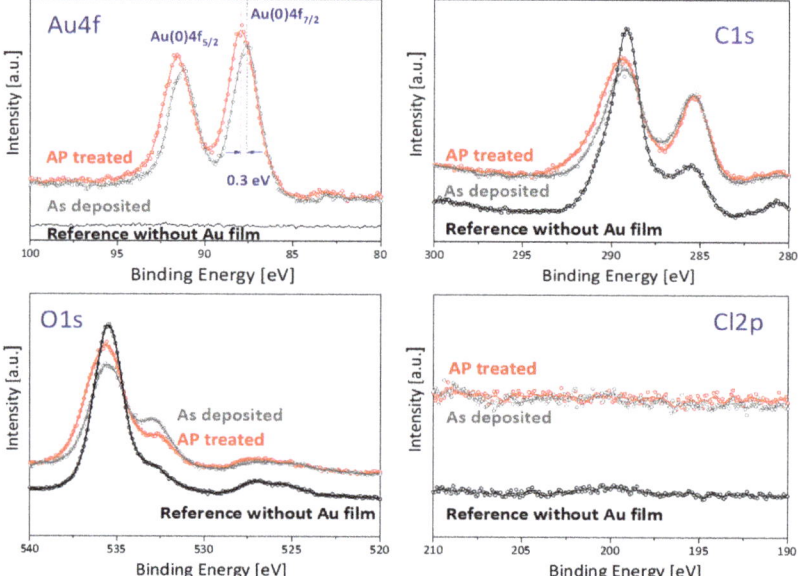

Figure 2. Elemental XPS spectra of Au4 f, C1 s, O1 s and Cl2 p for plasma-printed nanogold film before ('as deposited') and after ('AP treated') plasma post-treatment and the reference case of the paper base without gold film.

Table 1. Atomic composition of nanogold film deposited on paper, measured by XPS.

	C [at %]	O [at %]	Au [at %]	Cl [at %]
As deposited	63.9	34.1	2.0	0.0
Post-treated	59.6	38.3	2.1	0.0

Table 2 shows a comparison of the components of the C1 s peak for as-deposited and post-treated nanogold films with the components of the peak for the paper-based substrate without deposited gold. Casa XPS software was used to deconvolute the peaks with max. 0.5% error. The results indicate that plasma post-treatment has removed the amorphous carbon layer and increased mainly the amount of O-C=O bonds.

Table 2. Influence of plasma post-treatment on deposited nanogold XPS signals: comparison of components of C1s peak.

C1s.	Binding Energy [eV]/Composition [%]			
	C-C	C-O	C=O	O-C=O
Paper base	285.0 eV (21.1%)	286.7 eV (4.0%)	288.5 eV (65.5%)	290.2 eV (9.4%)
As deposited	284.8 eV (35.2%)	286.5 eV (5.7%)	288.6 eV (49.1%)	290.1 eV (10.1%)
Post-treated	284.8 eV (32.0%)	286.7 eV (2.8%)	288.7 eV (49.1%)	290.3 eV (16.1%)

3.3. Cocaine Analysis

Figure 3 shows the visible cocaine vibrational bands when tested on the plasma-deposited nanogold substrate after post-treatment, with increasing cocaine standard concentrations. It shows that the plasma-printed SERS substrate allows the detection of six to nine characteristic cocaine vibration bands. The bands and the corresponding vibration modes are listed in Table 3. Five bands, at 1003 cm^{-1}, 1032 cm^{-1}, 1164 cm^{-1}, 1450 cm^{-1} and 1600 cm^{-1}, were consistently enhanced across the concentrations tested. These bands correspond to the symmetric and asymmetric ring breathing, C-N stretching, asymmetric -CH$_3$ deformation and C=C aromatic ring stretching, respectively. Furthermore, at least one of the three C-C tropane ring stretching bands between 848–900 cm^{-1} was observed for each concentration. The band at 1200 cm^{-1}, corresponding to the other C-N stretching band, was observed at concentrations of 1, 100, 500 and 5000 ng/mL.

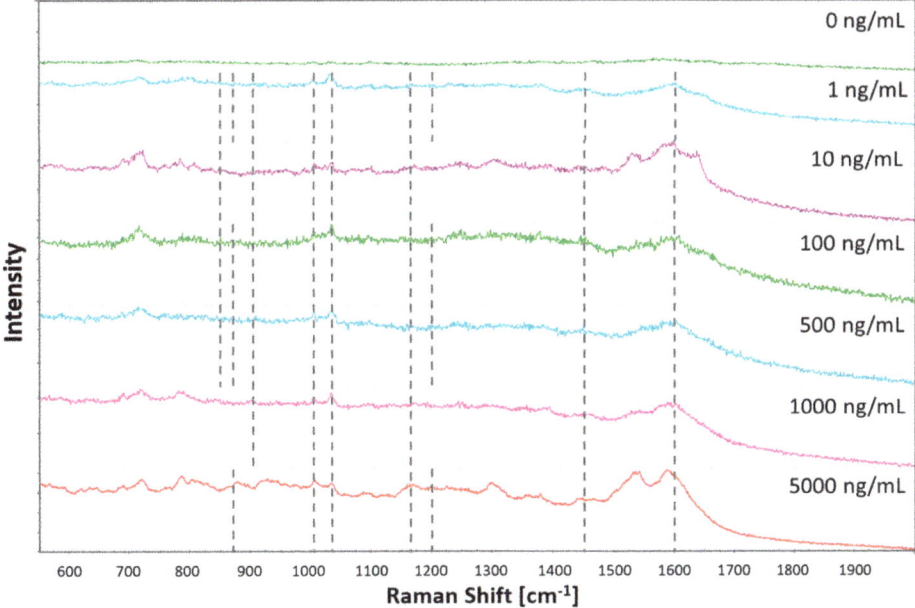

Figure 3. SERS spectra of cocaine standards of increasing concentration deposited onto plasma-printed substrates.

Table 3. Cocaine bands detected in cocaine standards on plasma-printed substrate.

Bands Listed in Ref. [3]		Plasma-Printed Substrate					
Vibration Mode	Cocaine HCl Salt	Cocaine HCl Salt	5000 ng/mL	1000 ng/mL	500 ng/mL	100 ng/mL	10 ng/mL	1 ng/mL
(C-C) stretching (tropane ring)	848 874 898	853 870 897	874		850 874 901	850 873 899	852	855 872 900
				898			898	
Symmetric stretching-aromatic ring breathing	1004	1001	1004	1005	1003	1004	1005	1004
Asymmetric stretching-aromatic ring breathing	1026	1027	1028	1027	1029	1026	1028	1027
C-N stretching	1165	1164	1168	1168	1164	1168	1170	1169
C-N stretching	1207	1205	1201	-	1208	1203	-	1202
Asymmetric CH$_3$ deformation	1462	1459	1455	1458	1454	1460	1455	1458
C=C stretching-aromatic ring	1596, 1601	1599	1595	1600	1601	1602	1599	1600
C=O symmetric stretching-carbonyl	1716	1717	-	-	-	-	-	-
C=O asymmetric stretching-carbonyl	1735	-	-	-	-	-	-	-

3.4. Spatial Distribution

The consistency of the enhancement across the plasma-printed substrate was determined using Raman mapping with four common cocaine band intensities. The intensity heat maps shown in Figure 4 have a good correlation of the intensities across the four bands. The distribution across the entire surface of the substrate was found to be consistent except along the edges where the deposited surface was no longer visible. The intensity across these four bands allows one to conclude that the hotspots were distributed across the surface resulting in surface-wide detection.

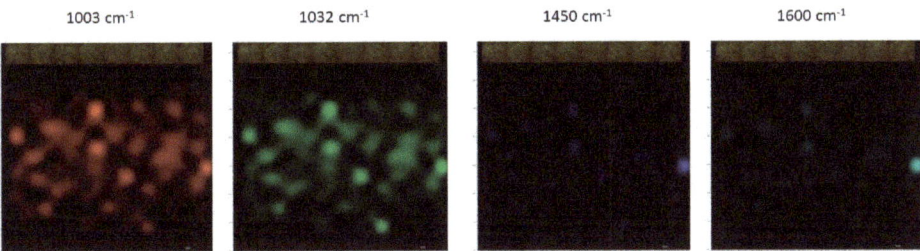

Figure 4. Raman heat maps showing the spatial distribution of cocaine using the four characteristic bands. The mapping steps were 175 μm × 175 μm in a grid for a total of 399 spectra collected across the surface area 3 mm × 3 mm.

3.5. Comparison with Commercial SERS Platform

The commercial substrates compared to the developed plasma printed substrate were a paper-based SERS (P-SERS) substrate with the gold SERS active metal deposited through inkjet printing and a silicon pillar-based substrate coated in silver (JASMAT Ag). These substrates were chosen for comparison as the P-SERS had a similar composition to the plasma-printed substrate and the JASMAT Ag had previously been shown to be effective for cocaine analysis [38].

Figure 5 shows Raman spectra measured on commercial P-SERS substrate with increasing cocaine standard concentrations. Only two Raman bands were consistently enhanced on the commercial substrate at ~1000 cm^{-1} and ~1027 cm^{-1} in the presence of cocaine. These correspond to the symmetric and asymmetric stretching of the aromatic ring. The band at ~1600 cm^{-1} from C=C aromatic stretching was observed for all the concentrations except the 10 ng/mL standard. The tropane and carbonyl bands were not observed in any of the samples. The C-N stretching band at ~1162 cm^{-1} was only observed once at 10 ng/mL concentration, while the second C-N stretching band at ~1198 cm^{-1} was only observed at a concentration of 1000 ng/mL and the asymmetric -CH$_3$ deformation band at ~1446 cm^{-1} was observed at 5000 ng/mL. When compared to the developed plasma deposited substrate results, this commercial substrate enhanced fewer cocaine vibrational bands at each concentration. Furthermore, only two bands were enhanced across the tested concentrations compared to five consistent bands on the developed substrate. For the detailed information on each different vibrational band detected on P-SERS substrate, see Table S1.

The commercial P-SERS substrate did have a more intense peak at both of the consistently enhanced bands. However, these bands are common among the drugs tested as they correspond to aromatic ring breathing bands. Therefore, vibrational bands need to be consistently enhanced to produce a characteristic fingerprint of the analyte. The analyte can only be confirmed if enough of the characteristic bands are visible.

The three substrates were compared using the number of bands enhanced at a concentration of 100 ng/mL as shown in Figure 6, and the number of bands consistently enhanced across the six concentrations, presented in the SI. The plasma-printed substrate enhanced between six and nine bands for each concentration. Five of these bands were consistently enhanced across all of the concentrations. The commercial P-SERS only enhanced three or four bands, with only two consistently enhanced. The commercial JASMAT Ag substrate enhanced between five and seven bands for cocaine, with four being consistently enhanced as shown in Figure S8 and Table S2 of SI. At the concentration of 100 ng/mL, shown in Figure 6, the plasma-printed substrate enhanced nine cocaine bands. The commercial P-SERS and JASMAT Ag substrates enhanced three and five bands, respectively. The cocaine bands tend to be more intense for the commercial substrates. The plasma-printed substrate outperformed the two commercial substrates for the analysis of cocaine based on both the number of enhanced bands and number of consistently enhanced bands.

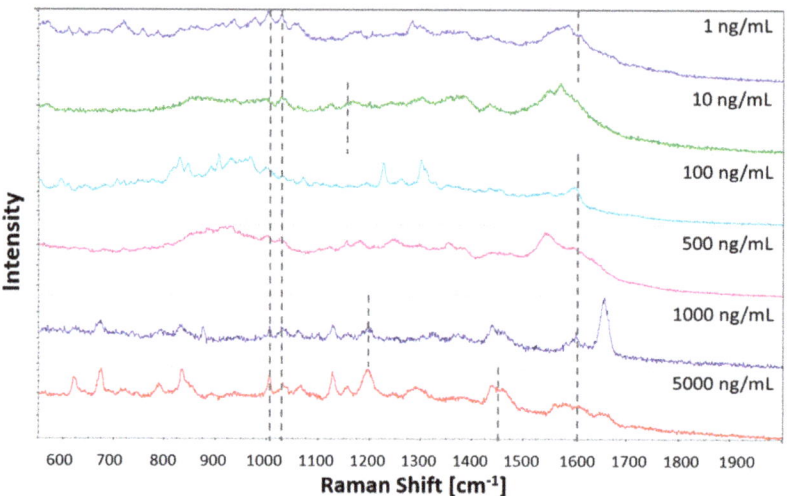

Figure 5. SERS spectra collected from cocaine deposited on commercial P-SERS substrate.

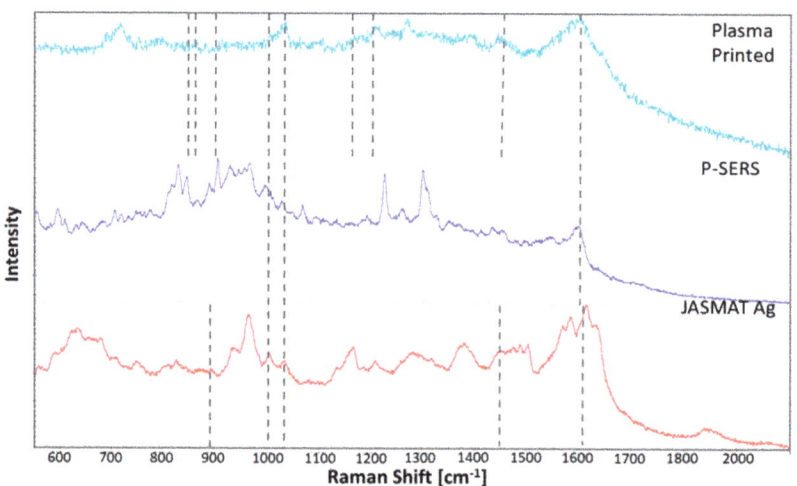

Figure 6. Comparison of cocaine at 100 ng/mL concentration deposited onto the plasma printed substrate and two commercial SERS substrates, P-SERS and JASMAT Ag.

3.6. Application to Oral Fluid

The plasma-printed substrate was tested with cocaine extracted from oral fluid spiked at cocaine concentrations of 10 ng/mL and 100 ng/mL. The results are presented in Figure 7. At 100 ng/mL, there were three visible cocaine bands. These correspond to (C-C) stretching of the tropane ring, symmetric aromatic ring breathing and C-N stretching. The lower concentration of 10 ng/mL revealed five cocaine bands. The enhanced bands corresponded to the (C-C) stretching of the tropane ring at 850 cm^{-1} and 897 cm^{-1}, symmetric aromatic ring breathing at 1003 cm^{-1}, and C-N stretching at 1164 cm^{-1} and 1205 cm^{-1}. Due to the possible interference from many other compounds and proteins in oral fluid, a lower number of enhanced bands was visible than in the standard of the same concentration. However, it is an important result to demonstrate the possible application of the plasma printed SERS strip in real application of on-site illicit drug testing.

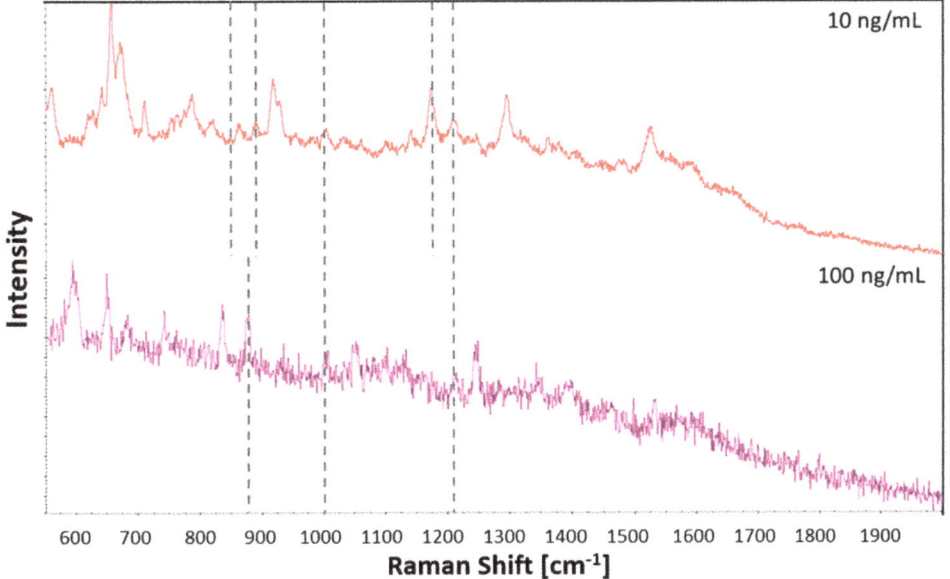

Figure 7. Comparison of SERS spectra acquired from extracted cocaine spiked oral fluid at 10 ng/mL and 100 ng/mL deposited onto the plasma printed substrate.

3.7. Cost Comparison

To demonstrate the cost-effectiveness of the plasma process, the production cost of the plasma-printed nanogold SERS substrate is estimated and compared to the price of the two commercial strips in Table S3 in the SI. The costs of electricity, gas, liquid precursor, filter paper and depreciation of the equipment such as power supply, liquid pump, mass flow meters and nebuliser are included based on the assumption of a five-year lifetime. The labour cost or other possible indirect expenses are not included. The total estimated cost to produce a single SERS substrate using the current plasma system is 0.107 AUD. This is doubled to take into account possible errors or interruptions in processing. The details of the cost estimation are given in Table S4 of the SI. For the cost calculation, the active area is presumed to be the same as that of the commercial JASMAT Ag substrate. However, the plasma-printing process enables continuous processing, unlike batch processes such as e-beam evaporation or sputtering, which also require an expensive high vacuum system. For large-scale processing, the current single jet system could be redesigned into an array or slit jet to cover a large area at the same time. With that modification, a further decrease in cost can be expected because liquid pumps and gas flow meters can be shared.

4. Discussion

Plasma-printed nanogold on a paper based was demonstrated to be a highly sensitive and cost-effective SERS substrate for cocaine detection. The plasma-printed substrate was able to detect between six and nine characteristic Raman peaks of cocaine at concentrations from 1 to 5000 ng/mL, whereas a commercial SERS gold on a paper-based substrate enhanced only three to four bands. In addition, the plasma-printed SERS substrate provided better consistency than the commercial SERS substrate with Ag pillar structure, which is currently favoured due to its high sensitivity. It is likely that the direct plasma deposition of nanogold from a solution precursor provides desirable surface conditions without interference from residual chemicals, such as reducing or stabilising agents. The additional plasma post-treatment step removed the organic carbon layer, which may have formed as a result of the ethyl alcohol used as a diluting solvent to improve the atomisation of the precursor. The paper-based plasma-printed SERS substrate has the potential to be a practical, economical solution for on-site screening or point-of-care applications, such as illicit drug detection, when used in combination with a portable Raman system.

Supplementary Materials: The following are available online at http://www.mdpi.com/1424-8220/21/3/810/s1, Figure S1: Influence of the number of nanogold deposition passes on SERS performance, Figure S2: SEM images of plasma printed nanogold film on paper substrate for SERS measurement, Figure S3: Comparison of SERS spectra of nanogold substrates with and without post-treatment, Figure S4: Comparison of SERS spectra of nanogold substrates with and without post-treatment using 10^{-6} M of Rhodamine B aqueous solution, Figure S5: Optical emission spectra in the range of 200–800 nm, Figure S6: Optical emission from different species in the active discharge region, Figure S7: Optical emission from different species near the substrate and Figure S8: Comparison of SERS spectra acquired from standard cocaine solutions with decreasing concentrations deposited onto commercial JASMAT Ag; Table S1: Cocaine bands detected in cocaine standards on the commercial P-SERS substrate, Table S2: Cocaine bands detected in cocaine standards on the JASMAT Ag substrate, Table S3: Comparison of SERS substrates showing nanoparticle type, backing substrate, deposition technique, particle size, size of active area and cost per substrate and Table S4: Estimated cost of SERS strip printing using plasma jet based on current lab-scale system.

Author Contributions: Conceptualisation, K.O., S.F. and A.B.M.; methodology, F.I., L.X., B.A., C.C. and A.B.; validation, F.I., L.X. and J.F.; formal analysis, R.A., J.H., C.C., A.B. and J.F.; investigation, R.A. and J.H.; resources, A.B.; data curation, E.C. and J.F.; writing—original draft preparation, R.A. and J.H.; writing—review and editing, E.C., B.A. and A.B.M.; visualization, R.A. and J.H.; supervision, K.O., S.F. and A.B.M.; project administration, S.F. and A.B.M.; funding acquisition, K.O., S.F. and A.B.M. All authors have read and agreed to the published version of the manuscript.

Funding: This research is supported by an Australian Government Research Training Program Scholarship awarded to R.A. and the Integrated Device for End-User Analysis at Low Levels ARC Research Hub.

Institutional Review Board Statement: Not applicable.

Informed Consent Statement: Not applicable.

Data Availability Statement: Data sharing not applicable.

Conflicts of Interest: The authors declare no conflict of interest.

References

1. Haddad, A.; Comanescu, M.A.; Green, O.; Kubic, T.A.; Lombardi, J.R. Detection and quantitation of trace fentanyl in heroin by surface-enhanced raman spectroscopy. *Anal. Chem.* **2018**, *90*, 12678–12685. [CrossRef] [PubMed]
2. Leonard, J.; Haddad, A.; Green, O.; Birke Ronald, L.; Kubic, T.; Kocak, A.; Lombardi, J.R. SERS, Raman, and DFT analyses of fentanyl and carfentanil: Toward detection of trace samples. *J. Raman Spectrosc.* **2017**, *48*, 1323–1329. [CrossRef]
3. de Oliveira Penido, C.A.F.; Pacheco, M.T.T.; Lednev, I.K.; Silveira, L. Raman spectroscopy in forensic analysis: Identification of cocaine and other illegal drugs of abuse. *J. Raman Spectrosc.* **2016**, *47*, 28–38. [CrossRef]

4. Andreou, C.; Hoonejani, M.R.; Barmi, M.R.; Moskovits, M.; Meinhart, C.D. Rapid detection of drugs of abuse in saliva using surface enhanced raman spectroscopy and microfluidics. *ACS Nano* **2013**, *7*, 7157–7164. [CrossRef] [PubMed]
5. Rana, V.; Cañamares, M.V.; Kubic, T.; Leona, M.; Lombardi, J.R. Surface-Enhanced raman spectroscopy for trace identification of controlled substances: Morphine, codeine, and hydrocodone. *J. Forensic Sci.* **2011**, *56*, 200–207. [CrossRef] [PubMed]
6. Botti, S.; Almaviva, S.; Cantarini, L.; Palucci, A.; Puiu, A.; Rufoloni, A. Trace level detection and identification of nitro-based explosives by surface-enhanced Raman spectroscopy. *J. Raman Spectrosc.* **2013**, *44*, 463–468. [CrossRef]
7. Hakonen, A.; Wang, F.; Andersson, P.O.; Wingfors, H.; Rindzevicius, T.; Schmidt, M.S.; Soma, V.R.; Xu, S.; Li, Y.; Boisen, A.; et al. Hand-Held femtogram detection of hazardous picric acid with hydrophobic Ag nanopillar SERS substrates and mechanism of elasto-capillarity. *ACS Sens.* **2017**, *2*, 198–202. [CrossRef]
8. Strelau, K.K.; Schüler, T.; Möller, R.; Fritzsche, W.; Popp, J. Novel bottom-up SERS substrates for quantitative and parallelized analytics. *ChemPhysChem* **2010**, *11*, 394–398. [CrossRef]
9. Moskovits, M. Surface-Enhanced spectroscopy. *Rev. Mod. Phys.* **1985**, *57*, 783–826. [CrossRef]
10. Fedick, P.W.; Morato, N.M.; Pu, F.; Cooks, R.G. Raman spectroscopy coupled with ambient ionization mass spectrometry: A forensic laboratory investigation into rapid and simple dual instrumental analysis techniques. *Int. J. Mass Spectrom.* **2020**, *452*, 116326. [CrossRef]
11. Yu, W.W.; White, I.M. Chromatographic separation and detection of target analytes from complex samples using inkjet printed SERS substrates. *Analyst* **2013**, *138*, 3679–3686. [CrossRef] [PubMed]
12. Song, C.Y.; Yang, B.Y.; Chen, W.Q.; Dou, Y.X.; Yang, Y.J.; Zhou, N.; Wang, L.H. Gold nanoflowers with tunable sheet-like petals: Facile synthesis, SERS performances and cell imaging. *J. Mater. Chem. B* **2016**, *4*, 7112–7118. [CrossRef] [PubMed]
13. He, S.; Chua, J.; Tan, E.K.M.; Kah, J.C.Y. Optimizing the SERS enhancement of a facile gold nanostar immobilized paper-based SERS substrate. *RSC Adv.* **2017**, *7*, 16264–16272. [CrossRef]
14. Li, J.P.; Zhou, J.; Jiang, T.; Wang, B.B.; Gu, M.; Petti, L.; Mormile, P. Controllable synthesis and SERS characteristics of hollow sea-urchin gold nanoparticles. *Phys. Chem. Chem. Phys.* **2014**, *16*, 25601–25608. [CrossRef] [PubMed]
15. Guo, P.Z.; Sikdar, D.; Huang, X.Q.; Si, K.J.; Xiong, W.; Gong, S.; Yap, L.W.; Premaratne, M.; Cheng, W.L. Plasmonic core-shell nanoparticles for SERS detection of the pesticide thiram: Size- and shape-dependent Raman enhancement. *Nanoscale* **2015**, *7*, 2862–2868. [CrossRef]
16. Felidj, N.; Aubard, J.; Levi, G.; Krenn, J.R.; Hohenau, A.; Schider, G.; Leitner, A.; Aussenegg, F.R. Optimized surface-enhanced Raman scattering on gold nanoparticle arrays. *Appl. Phys. Lett.* **2003**, *82*, 3095–3097. [CrossRef]
17. Zhao, X.Y.; Wen, J.H.; Zhang, M.N.; Wang, D.H.; Chen, Y.W.L.; Chen, L.; Zhang, Y.; Yang, J.; Dut, Y. Design of hybrid nanostructural arrays to manipulate SERS-Active substrates by nanosphere lithography. *ACS Appl. Mater. Interfaces* **2017**, *9*, 7710–7716. [CrossRef]
18. Sivashanmugan, K.; Liao, J.D.; You, J.W.; Wu, C.L. Focused-ion-beam-fabricated Au/Ag multilayered nanorod array as SERS-active substrate for virus strain detection. *Sens. Actuators B Chem.* **2013**, *181*, 361–367. [CrossRef]
19. Moram, S.B.; Byram, C.; Shibu, S.N.; Chilukamarri, B.M.; Soma, V.R. Ag/Au nanoparticle-loaded paper-based versatile surface-enhanced raman spectroscopy substrates for multiple explosives detection. *ACS Omega* **2018**, *3*, 8190–8201. [CrossRef]
20. Liana, D.D.; Raguse, B.; Wieczorek, L.; Baxter, G.R.; Chuah, K.; Gooding, J.J.; Chow, E. Sintered gold nanoparticles as an electrode material for paper-based electrochemical sensors. *RSC Adv.* **2013**, *3*, 8683–8691. [CrossRef]
21. Marques, A.; Veigas, B.; Araujo, A.; Pagara, B.; Baptista, P.V.; Aguas, H.; Martins, R.; Fortunato, E. Paper-Based SERS platform for one-step screening of tetracycline in milk. *Sci. Rep. UK* **2019**, *9*. [CrossRef] [PubMed]
22. Joshi, P.; Santhanam, V. Paper-based SERS active substrates on demand. *RSC Adv.* **2016**, *6*, 68545–68552. [CrossRef]
23. Lee, D.J.; Kim, D.Y. Hydrophobic paper-based SERS sensor using gold nanoparticles arranged on graphene oxide flakes. *Sensors* **2019**, *19*, 5471. [CrossRef] [PubMed]

24. Lee, C.H.; Tian, L.; Singamaneni, S. Paper-Based SERS swab for rapid trace detection on real-world surfaces. *ACS Appl. Mater. Interfaces* **2010**, *2*, 3429–3435. [CrossRef] [PubMed]
25. Yu, W.W.; White, I.M. Inkjet-printed paper-based SERS dipsticks and swabs for trace chemical detection. *Analyst* **2013**, *138*, 1020–1025. [CrossRef] [PubMed]
26. Yu, W.W.; White, I.M. Inkjet printed surface enhanced raman spectroscopy array on cellulose paper. *Anal. Chem.* **2010**, *82*, 9626–9630. [CrossRef]
27. Restaino, S.M.; White, I.M. Inkjet-Printed paper surface enhanced Raman spectroscopy (SERS) sensors Portable, low cost diagnostics for microRNA. *IEEE Sens.* **2016**, 1–3. [CrossRef]
28. Hoppmann, E.P.; Yu, W.W.; White, I.M. Inkjet-Printed fluidic paper devices for chemical and biological analytics using surface enhanced raman spectroscopy. *IEEE J. Sel. Top. Quantum* **2014**, *20*. [CrossRef]
29. Hoppmann, E.P.; Yu, W.W.; White, I.M. Highly sensitive and flexible inkjet printed SERS sensors on paper. *Methods* **2013**, *63*, 219–224. [CrossRef]
30. Hong, J.M.; Yick, S.; Chow, E.; Murdock, A.; Fang, J.H.; Seo, D.H.; Wolff, A.; Han, Z.J.; van der Laan, T.; Bendavid, A.; et al. Direct plasma printing of nano-gold from an inorganic precursor. *J. Mater. Chem. C* **2019**, *7*, 6369–6374. [CrossRef]
31. Australian Criminal Intelligence Commission. *National Wastewater Drug Monitoring Program—Report 6*; Australian Criminal Intelligence Commission: Canberra, Australia, 2019.
32. Australian Criminal Intelligence Commission. *Illicit Drug Data Report 2017–2018*; Australian Criminal Intelligence Commission: Canberra, Australia, 2019.
33. Baik, J.H. Dopamine signaling in reward-related behaviors. *Front. Neural Circuits* **2013**, *7*, 152. [CrossRef] [PubMed]
34. Brands, B.; Sproule, B.; Marshman, J. *Drugs & Drug Abuse*, 3rd ed.; Centre for Addiction and Mental Health: Toronto, ON, Canada, 1998; p. 658.
35. Brunton, L.; Knollman, B.; Chabner, B. *Goodman & Gilman's: The Pharmacological Basis of Therapeutics*; McGraw-Hill Medical: New York, NY, USA, 2011.
36. Julien, R.M.; Advokat, C.D.; Comaty, J.E. *Primer of Drug Action*; Worth Publishers: New York, NY, USA, 2010.
37. Baldock, M.R.J.; Wooley, J.E. Reviews of the effectiveness of random drug testing in Australia: The absence of crash-based evaluations. In Proceedings of the Australasian Road Safety Research, Policing & Education Conference, Brisbane, Australia, 28–30 August 2013.
38. Alder, R.; Xiao, L.D.; Fu, S.L. Comparison of commercial surface-enhanced Raman spectroscopy substrates for the analysis of cocaine. *Drug Test. Anal.* **2020**, 1–9. [CrossRef]
39. Clauwaert, K.M.; Van Bocxlaer, J.F.; Lambert, W.E.; De Leenheer, A.P. Liquid chromatographic determination of cocaine, benzoylecgonine, and cocaethylene in whole blood and serum samples with diode-array detection. *J. Chromatogr. Sci.* **1997**, *35*, 321–328. [CrossRef] [PubMed]
40. Wu, T.-J.; Chou, C.-Y.; Hsu, C.-M.; Hsu, C.-C.; Chen, J.-Z.; Cheng, I.C. Ultrafast synthesis of continuous Au thin films from chloroauric acid solution using an atmospheric pressure plasma jet. *RSC Adv.* **2015**, *5*, 99654–99657. [CrossRef]
41. Maguire, P.; Rutherford, D.; Macias-Montero, M.; Mahony, C.; Kelsey, C.; Tweedie, M.; Pérez-Martin, F.; McQuaid, H.; Diver, D.; Mariotti, D. Continuous in-flight synthesis for on-demand delivery of ligand-free colloidal gold nanoparticles. *Nano Lett.* **2017**, *17*, 1336–1343. [CrossRef]
42. Boudou, J.P.; Paredes, J.I.; Cuesta, A.; Martinez-Alonso, A.; Tascon, J.M.D. Oxygen plasma modification of pitch-based isotropic carbon fibres. *Carbon* **2003**, *41*, 41–56. [CrossRef]
43. Kan, C.W.; Man, W.S. Surface characterisation of atmospheric pressure plasma treated cotton fabric-effect of operation parameters. *Polymers* **2018**, *10*, 250. [CrossRef]

Publisher's Note: MDPI stays neutral with regard to jurisdictional claims in published maps and institutional affiliations.

© 2021 by the authors. Licensee MDPI, Basel, Switzerland. This article is an open access article distributed under the terms and conditions of the Creative Commons Attribution (CC BY) license (http://creativecommons.org/licenses/by/4.0/).

MDPI
St. Alban-Anlage 66
4052 Basel
Switzerland
Tel. +41 61 683 77 34
Fax +41 61 302 89 18
www.mdpi.com

Sensors Editorial Office
E-mail: sensors@mdpi.com
www.mdpi.com/journal/sensors

www.ingramcontent.com/pod-product-compliance
Lightning Source LLC
LaVergne TN
LVHW070604100526
838202LV00012B/557